THE BULLETIN
Volume 35   1996

Mount Desert Island
Biological Laboratory
Salsbury Cove, Maine 04672

TABLE OF CONTENTS

THE MOUNT DESERT ISLAND BIOLOGICAL LABORATORY

SERVING HUMANITY THROUGH
RESEARCH AND TEACHING IN MARINE BIOMEDICINE

INTRODUCTION

The Mount Desert Island Biological Laboratory (MDIBL) is an independent non-profit biological station. It is located on the north shore of Mount Desert Island, overlooking the Gulf of Maine about 120 miles northeast of Portland near the mouth of the Bay of Fundy. The island, well known for Acadia National Park, provides a variety of habitats including shallow and deep saltwater, a broad intertidal zone, saltwater and freshwater marshes, freshwater lakes and streams, forests and meadows.

The Laboratory is the largest cold water research facility in the Eastern United States, and its unique site provides an outstanding environment for studying the physiology of marine and freshwater flora and fauna. During 1995, the scientific personnel included 49 principal investigators, 70 associates and 11 assistants/technicians, representing 51 institutions in 25 states and 5 European countries.

HISTORY AND ORGANIZATION

MDIBL was founded in 1898 at South Harpswell, Maine by J.S. Kingsley of Tufts University. Its present site at Salsbury Cove was donated by the Wild Gardens of Acadia and relocation was completed in 1921. The Wild Gardens of Acadia, a land-holding group headed by George B. Dorr and John D. Rockefeller, Jr., was instrumental in the founding of Acadia National Park.

The Laboratory was incorporated in 1914 under the laws of the State of Maine as a non-profit scientific and educational institution. Founded as a teaching laboratory, MDIBL is now a center for marine biomedical research and teaching that attracts investigators and students from across the U.S. and around the world. Since the pioneering work of H.W. Smith, E.K. Marshall, and Roy P. Forster on various aspects of renal and osmoregulatory physiology of local fauna, the Laboratory has become known worldwide as a center for investigations in electrolyte and transport physiology, developmental biology and electrophysiology.

The Mount Desert Island Biological Laboratory is owned and operated by the Board of Trustees and Members of the Corporation; at present, there are 434 members. Officers of the Corporation - Chair, Vice-Chair, Director, Secretary, Treasurer, Clerk - and an Executive Committee are elected from among the Trustees. The Chair and Executive Committee oversee the general administration and long range goals of the Laboratory. The Director, with the aid of a full-time Administrative Director and staff, is responsible for implementing the scientific, educational and public service activities of the Laboratory.

NIEHS TOXICOLOGY CENTER

In 1985, with the support of the National Institute of Environmental Health Sciences (NIEHS), MDIBL established a center dedicated to the study of the toxic effects of heavy metals and other environmental pollutants that pose an increasing health risk to humans and a threat to the marine environment. The focus of The Center for Membrane Toxicity Studies (CMTS), is the use of marine animals like the shark, the flounder and the skate to define sites of action for metals such as mercury and cadmium that enter the environment due to improper disposal of industrial waste and as a component of some pesticides. The effects of these pollutants are wide-spread in the human body, with affected organs including the brain, the kidney, the liver, the gastrointestinal tract and the reproductive system. The goal of the CMTS is to identify the molecular targets for toxic substances and to provide the scientific basis for the development of treatments for heavy-metal intoxication. Inquiries concerning the center are welcome.

APPLICATIONS & FELLOWSHIPS

Research space is available for the entire summer season (June 1 - September 30) or a half-season (June 1 - July 31 or August 1 - September 30). Applications for the coming summer must be submitted by February 1st each year. Investigators are invited to use the year-round facilities at other times of the year, but such plans should include prior consultation with the MDIBL Office concerning available facilities and specimen supply.

A number of fellowships and scholarships are available to research scientists, undergraduate faculty and students, and high school students. These funds may be used to cover the cost of laboratory rent, housing and supplies. Stipends are granted with many of the student awards. Applications for fellowships for the coming summer research period are generally due in January.

For further information on applications and fellowships/scholarships, please contact:

Dr. Barbara Kent
Mount Desert Island Biological Laboratory
P.O. Box 35
Salsbury Cove, ME 04672
Tel. (207) 288-3605
Fax. (207) 288-2130
e-mail: bkb@mdibl.org

ACKNOWLEDGEMENTS

The Mount Desert Island Biological Laboratory is indebted to the National Science Foundation and National Institutes of Health for substantial support. Funds for building renovations and new construction continue to permit the Laboratory to expand and upgrade its research and teaching facilities. Individual research projects served by the Laboratory are funded by private and government agencies, and all of these projects have benefited from the NSF and NIH grants to the Laboratory. For supporting our educational initiative, MDIBL acknowledges the Burroughs Wellcome Fund, Grass Foundation, Milbury Fellowship Fund, American Heart Association - Maine Affiliate, Mr. Robert E. Blum, Maine Community Foundation, NSF - Research Experience for Undergraduates and NSF Young Scholar Program for High School students and many local businesses and individuals.

THE HOMER W. SMITH SYMPOSIUM 1995: REFLECTIONS AND IMPRESSIONS

Rolf K.H. Kinne and E. Kinne-Saffran
Max-Planck-Institut für molekulare Physiologie,
44139 Dortmund, FRG

Preface

On the occasion of the 100th anniversary of his birth an international symposium honouring Homer W. Smith was held on Mount Desert Island under the auspices of the Mount Desert Island Biological Laboratory from August 15 till August 19, 1995. The idea to organize such a symposium originated from Yuri Natochin, who discussed it with Klaus Beyenbach, who in turn got the Kinne's and the MDIBL involved. The stone that had been thrown in the water created ever growing circles of interest that soon reached Barry Brenner, who added the aspects of clinical nephrology to the program and secured the funding of the symposium.

But how does one honour appropriately such an eminent man, who excelled as "Scientist, Teacher, Explorer, Novelist, and Perennial Student" [S.J. Farber, Kidney Int. 49, 1996]. As scientists the organizers decided to dedicate most of the symposium to evaluate the progress made in renal physiology and clinical nephrology since Homer W. Smith deceased in 1962. As a guideline we selected his pivotal monograph: The Kidney: Structure and Function in Health and Disease, published in 1951 [Oxford University Press, New York] and used its title for the symposium.

Honouring Homer W. Smith meant not only to recall his role in nephrology "a field that he dominated for over thirty years in a way few (if any) have dominated other fields" [R.F. Pitts, National Acad. Sci. 39, 445-470, 1967], but also to pay tribute to some of the other prominent aspects of Smith's multifacetted personality and interests. One of these was his influence on others: "In fact, I believe it would be fair to say, that Smith's contributions to renal physiology and clinical nephrology should be measured even more by his influence on others and their productivity than by his own scientific achievements, great as they were. A generation of renal physiologists and clinical nephrologists owe their expertise to the direct influence of Homer W. Smith" [D.S. Baldwin, Kidney Int. 49, 1996]. Such influence did manifest itself in several ways. For many participants at the symposium his 1937 book on the Physiology of the Kidney [Oxford University Press, New York] was their fist significant exposure to the subject, "and provided the real impetus for them to have a try at that field" because observations of various origins had been synthesized into a "logical and consistent picture of how the kidney works" [R.W. Berliner, Kidney Int. 49, 1996]. Homer W. Smith had also a deep

interest in "mentorship at various levels of sophistication...
and a very special concern for and guidance of his mentees"
[I.L. Schwartz, Kidney Int. 49, 1996]. This tutelage reached
from individual mentor-mentee relationships to medical education
in general.

Therefore, in order to get (and provide the participants of
the symposium with) a glance of Homer W. Smith as a human being,
at the opening event disciples and acquaintances of Homer W.
Smith were asked to share their experiences and impressions and
set the tone of the symposium to come.

In addition, a Round Table Discussion was incorporated into
the symposium, entitled 'Nephrology in 1995: More and More
Details, Less and Less Synthesis. Is this the Smith Legacy?', in
which the current concerns about a proper balance between the
reductionistic and holistic approach to renal physiology and
medical education was debated.

A third area of Smith's interest was comparative physiology
and evolution. To cover this topic a special lecture was arranged, and we were lucky to recruit Stephen Jay Gould as a speaker.

And last, but not least, Smith's instrumental role in the
scientific life of the MDIBL was highlighted by an exhibit
'Homer W. Smith at Work at the MDIBL', arranged by Carl W.
Gottschalk.

If one leans back and asks oneself what do you remember most
of the symposium it is ... the weather. We had a perfect week in
August with bright sunshine, which bathed the short commute, the
coffee breaks and the receptions in brilliance and beauty. And
we had the breeze, moving the shades in the auditorium from time
to time as if to remind us of the world outside the ivory tower
we were contemplating in, and like in a sail boat filling the
sails of science to reach new frontiers. Then the participants
come to mind - in their cheerful mood they created the spirit of
the symposium and lived it. Some of them had not seen each other
for years, others - especially the more junior scientists - met
for the first time men and women who shaped the field of nephrology for the last decades. And of course the scientific presen-
tations, thoughtfully prepared and masterfully presented.

In the following we will share with you our reflections on
the outcome of this endeavor and highlight some of the recent
developments in those scientific areas Homer W. Smith was and
investigators at MDIBL currently are interested in. The account
is very personal but we think that we can afford this privilege

v

because the Proceedings of the Symposium will be published in a forthcoming issue of Kidney International this summer[1].

Excretion of Urea*

Until recently, our thinking about the handling of urea by cells and in the renal medulla was governed by the assumption that urea easily crosses cell membranes by simple lipid diffusion and that its movements are governed by passive transport across the epithelia. Both assumptions have been challenged in recent years. Sands et al. [ibid] provided clear evidence that in rats, when fed a low protein diet, the initial inner medullary collecting duct, when perfused in vitro, transports urea actively. Since removal of sodium from the tubular lumen or addition of ouabain to the basal-lateral cell side reduces transport significantly, it is postulated that the transport is secondary active involving a sodium-urea cotransport system in the luminal membrane.

As reviewed by Hediger et al. erythrocytes and epithelial cells of certain nephron segments have urea permeabilities that are considerably higher than expected from simple diffusion, suggesting the presence of special urea transporters. Some of these transporters have now been cloned and identified at the molecular level. Expression cloning from rabbit and subsequently from rat kidney has led to the identification of a novel phloretin-sensitive urea transporter, called UT2. The sequence predicts a 43 kDa polypeptide with ten transmembrane domains and one predominant extracellular loop between transmembrane helix 5 and 6. Rat UT2 has two transcripts in the kidney, one of 2.9 kb and another of 4.0 kb that show a different distribution in the kidney. The 2.9 transcript is found in highest concentrations in the inner stripe of the outer medulla and the inner medulla, whereas the 4.0 kb transcript is present mainly in the inner medulla and papilla of the rat. In addition, the two transcripts respond differently to various physiological states of the animals. The 4.0 transcript is regulated by the dietary protein uptake and the 2.9 kb transcript by the hydration state of the animal. Thus, tools are now available to further characterize the molecular and cellular events underlying the regulation of urea transport in the medulla. The information at the molecular level will undoubtedly also help to further clarify the complex pathways and mechanisms of urea recycling that have been brought into focus again by Bankir et al. [ibid].

---

[1] All articles cited in the following will be published in Kidney International, Vol. 49, 1996 as Proceedings of the Homer W. Smith Symposium
* Indicates a title of a chapter in The Kidney: Structure and Function in Health and Disease

Clearances Involving Active Tubular Reabsorption[*]

The field of tubular transport is currently dominated by two major themes: the molecular identification of the pumps, channels and transporters involved in tubular transport and the cellular events regulating the activity of transport systems either in loco or by sorting to the appropriate membrane domains. Establishment and maintenance of the biochemically, structurally and physiologically distinct apical and basolateral domains of the plasma membranes of epithelial cells thereby is of central importance for the proper function of these cells. As demonstrated by Molitoris and Wagner alterations in this process can be the basis for malfunction. Ischemia, via intracellular ATP-depletion, leads for example to a loss of proximal tubule cell surface membrane polarity. Of major importance for this process to take place is a rapidly occurring, duration-dependent disruption and dissociation of the actin cytoskeleton and associated surface membrane structures. This results in loss of cell-cell contact, cell-extracellular matrix adhesion and surface membrane polarity. A distribution of surface membrane proteins and lipids into alternate domains of the plasma membrane ensues with severe impairment of the vectorial transepithelial transport. These changes are reversible and the repair is probably supported by growth factors. The extracellular matrix has also been shown to be relevant, in addition to soluble mediator substances, in determining the proliferative and synthetic type of the glomerular mesangial cells. Changes in the extracellular matrix therefore may have a pivotal role in the altered behavior of mesangial cells in glomerular diseases of the kidney [Ruprecht et al.; Schlöndorff, ibid].

Clearances Involving Tubular Excretion[*]

The demonstration of tubular secretion in the kidney is strongly associated with E.K. Marshall, the MDIBL and the power of comparative physiology. With regard to the cellular mechanisms involved in the active secretion of organic anions Pritchard and Miller reported on a novel mechanism. According to their recent findings a significant fraction of organic anions having entered the cell via the dicarboxylate-organic anion exchanger is sequestered into vesicles. Disruption of the cellular microtubular network can lead to both, diminished vesicular movement within the cell and reduced transepithelial transport. Thus, vesicular transport appears to play a much more significant role in organic anion secretion than previously assumed. Whether the substantial intracellular sequestration of organic cations is also related to transepithelial secretion remains to be determined.

---

[*] Indicates a title of a chapter in The Kidney: Structure and Function in Health and Disease

The Antidiuretic Hormone and the Excretion of Water*

The understanding of the action of antidiuretic hormone on the water permeability in the collecting duct and the basis for the high hydraulic conductivities in the proximal tubule, the thin descending limb of Henle's loop and the collecting duct (in the presence of ADH) has increased steadily over the last decades and has now reached the molecular level. The new tools of cell biology and molecular biology do not only allow to describe in more detail the physiological events of water transfer across epithelia and its regulation by ADH but also to gain insight into the molecular basis of nephrogenic diabetes insipidus.

As reviewed by Knepper et al. and Nielsen et al. [ibid] the high hydraulic conductivity of certain tubular structures can be explained by the presence of water channels or 'aquaporins' in the plasma membranes of these cells. There are several aquaporins that differ in their regional distribution, cellular location and regulatory response. AQP1 is constitutively expressed at very high levels in the proximal tubule and in the descending limb of Henle's loop and present in both the apical and basal-lateral membranes of the cells, suggesting that rapid water transport across these epithelia is mediated by this water channel at both cell sides. AQP2 is found predominantly in the apical plasma membrane and in subapical membrane vesicles within principal cells of the collecting duct in cortex, outer medulla, and inner medulla. AQP2 is also the 'vasopressin-regulated water channel' that, as response to a stimulus by ADH, is incorporated into the luminal membrane by the 'shuttle mechanism'. These vesicles seem to be guided to the plasma membrane by 'vesicle-associated membrane proteins' such as synaptobrevins and by target membrane-associated proteins, such as syntaxin as demonstrated also by Hays [ibid]. Thus, the principles of cytoskeletal control and vesicle docking found in nerve terminals or the chromaffin cell seem to apply also to the collecting duct cells. AQP2 is also the target of long-term regulation of water permeability of inner medullary collecting duct (IMCD) cells. For example, restriction of fluid intake in rats for 24 hours leads to a marked increase of AQP2 expression in the apical membrane as well as the subapical membrane vesicles. Concomitantly, the corresponding mRNA was found to be augmented. AQP3 and AQP4 are present in the basolateral membranes of collecting duct principal cells and IMCD cells, but not elsewhere in the kidney. AQP3 is dominant in the lateral plasma membrane, whereas AQP4 is distributed almost equally between the basal and the lateral membranes. AQP2 mutations have been found in humans with primary nephrogenic diabetes insipidus and the lack of appropriate expression of AQP2 has been invoked in acquired forms of nephrogenic diabetes insipidus from a variety of causes.

---

* Indicates a title of a chapter in The Kidney: Structure and Function in Health and Disease

Vasopressin receptors that initiate the chain of events comprising the action of ADH on the water permeability of the collecting duct have also been cloned recently and have been studied in great detail with regard to their role in nephrogenic diabetes insipidus. As reviewed by Bichet, the human V2 receptor gene is located in chromosome region Xq28 and has three exons and two small introns. The cDNA sequence predicts a polypeptide of 371 amino acids with a structure typical for G-protein-coupled receptors with seven transmembrane, four extracellular and four cytoplasmic domains. As reported by Bichet [ibid], 68 putative disease-causing mutations in the gene of the human V2 receptor have now been reported in 95 presumably unrelated families with x-linked nephrogenic diabetes insipidus. These mutations in most cases can lead to one of three phenotypes: (1) impairment of hormone binding to the receptor at the cell surface, (2) blocked intracellular transport of the receptor, or (3) ineffective biosynthesis and/or accelerated degradation of the receptor. These studies are prime examples "to allay the concern ... that renal physiologists have abandonned the sick to pursue complicated and specious theorems in ivory towers, or have overlooked ... that in this molecular world, structure always underlies function [H.W. Smith, The Kidney: Structure and Function in Health and Disease].

Excretion of Sodium and Other Strong Electrolytes*

Homer Smith in his 1937 book [Physiology of the Kidney] wrote: "The history of renal physiology has erred, more often than not, by attempts at oversimplification. The problems of water and salt excretion appear to be extremely complex, and especially liable to this danger". The papers presented at the symposium confirm this complexity. Blaustein discussed the problem how impairment of sodium excretion - the major culprit in the pathogenesis of hypertension - leads to increased vascular resistance and elevation of the blood pressure. He invoked a recently discovered adrenal cortical hormone, 'endogenous oubain', as the mediator. This hormone inhibits the sodium pump and raises intracellular sodium in many cells. Due to a decrease in Na/Ca exchange the intracellular calcium and, more importantly, the capacity of intracellular calcium stores are increased. Consequently, the vascular smooth muscles, vasomotor neurons, and endothelial cells become hypersensitive to activation. This may account for the increased arterial tone and peripheral vascular resistance. Aperia et al. also proposed that the Na,K-ATPase is a major site of regulation for sodium balance. Her group demonstrated that the Na,K-ATPase protein is subject to an intricate system of phosphorylation (leading to inhibition) and dephosphorylation (leading to reactivation). These

---

* Indicates a title of a chapter in The Kidney: Structure and Function in Health and Disease

reactions are mastered by hormones such as dopamine and nor-epinephrine, which exert opposing forces on a common intracellular signalling system. This model also applies to the action of the natriuretic factor and its second messenger cGMP as well as to the antidiuretic factors, angiotensin and neuropeptide Y.

Na,K-ATPase has been the first sodium transport system to be identified in molecular terms and thus could be studied extensively with regard to its role in sodium homeostasis. Other sodium transport systems have followed suit recently. Aronson reported on the identification of the molecular entity of the Na/H exchanger in the proximal tubule that is responsible - in conjunction with the chloride-formate and chloride-oxalate exchangers - for the bulk of both, $NaHCO_3$ and NaCl reabsorption. There are several isoforms of Na/H exchangers that have been cloned. Out of these NHE1 appears to be located only in the basal-lateral plasma membranes of renal cells and is probably mainly involved in the regulation of intracellular pH. NHE3, on the other hand, in rat and rabbit kidney is predominantly localized in the brush border membrane of the proximal tubule as well as in subapical membrane vesicles. Apical membranes of other renal cells show only weak staining for NHE3. From functional studies the author concludes that "virtually all Na/H exchange activity in the renal brush border membranes is mediated by NHE3 under baseline conditions and during upregulation ... associated with metabolic acidosis, renal maturation, and glucocorticoid administration".

Also the major sodium transporter in the thick ascending limb and the distal tubule have now been identified. As reviewed by Hebert et al. the Na-K-2Cl cotransporter in the luminal membrane of the thick ascending limb of Henle's loop and the NaCl cotransporter in the early distal tubule evolved from a common ancestral gene and form a new gene family. The electroneutral sodium-chloride transport mechanism was first cloned from flounder urinary bladder, one of the model systems extensively used at the MDIBL, and isoforms have been found in a variety of mammalian species. Due to its sensitivity to thiazides it is called TSC (thiazide-sensitive sodium-chloride transporter). Bumetanide-sensitive Na-K-2Cl cotransporters, termed BSC, were identified thereafter. In the mouse BSC1 and BSC2 are the products of different genes and also exhibit different cellular locations and cellular function. BSC1 seems to be responsible for sodium-chloride reabsorption via secondary active chloride transport in the thick ascending limb of Henle's loop. BSC2 mediates Na-K-2Cl cotransport in secretory epithelia such as the rectal gland of the shark and is involved in volume regulation during cell shrinkage in a variety of tissues.

Finally, Benos et al. summarized the current status of knowledge on amiloride-sensitive sodium channels that modulate to a fine degree the final composition of the urine and thus are essential for sodium homeostasis. The unique feature of these channels is the wide variety of regulatory signals to which they respond. The activity of the channels is regulated by vasopressin, aldosterone, insulin, and atrial natriuretic factor. In some cases these hormonal effects are not only additive, but synergistic. Potential intracellular modulators, that have been identified very clearly in experiments employing purified proteins reconstituted in planar lipid bilayers, include protein kinase A, protein kinase C, tyrosine kinase, G- proteins, leukotrienes, cytoskeletal interactions and carboxymethylation reactions. The authors propose that the recently cloned epithelial sodium channel (ENaC) forms the central conducting element of all classes of amiloride-sensitive sodium channels, and that differences in kinetic parameters, ion selectivity, and sensitivity to regulatory factors are due to regional and tissue-specific modulators that associate with ENaC to form heteroisomers differing in stoichiometry and nature of components.

Volume Regulation and Organic Osmolytes in Renal Medullary Cells

As stated so pertinently by Burg in his contribution to the symposium "Homer Smith investigated marine organisms in order to understand better the functions of mammalian kidneys. The wisdom of this approach has been confirmed repeatedly and is evident in any consideration of compatible organic osmolytes in renal medulla". However, organic osmolytes or cell volume regulation are among the few items not mentioned in Smith's monograph The Kidney. To date five organic osmolytes have been identified in renal medullary cells: inositol, sorbitol, glycerophosphorylcholine (GPC), betaine, and taurine. Sorbitol and GPC are synthesized within the cells; for the former the rate of synthesis and for the latter the rate of degradation is controlled by extracellular osmolality. Inositol, betaine, and taurine are taken up by the cells via specific transport mechanisms whose activity is upregulated during hypertonicity of the extracellular medium. The extent of accumulation thereby is postulated to be controlled by the total concentration of all organic osmolytes - irrespective of the concentration of the individual components. The common signal being the intracellular ionic strength which apparently acts as signal for the accumulation [M. Burg, ibid]. According to differential display studies by Gullans et al., renal epithelial cells exposed to hyperosmolar sodium-chloride show an increased transcription of transporters, stress proteins, and metabolic enzymes. In contrast, urea initiates a different and very specific program that appears to involve a urea sensor which activates transcription and translation of the transcription factor Egr-1. Handler and Kwon reported about the identification of a tonicity responsive element (TonE) on the gene for the canine betaine cotransporter that consists of 13 base pairs and is localized upstream to the first exon. It conveys a hypertonicity-inducible increase in transcription to its own or a heterologous promotor in MDCK cells. They also identified a more acute regulation of the transport activity for inositol and betaine. This regulation is probably posttranslational and involves protein kinase A and C. Activation of both enzymes inhibits uptake of the osmolytes by about 30%.

Organic osmolytes are not only instrumental for the survival and maintenance of the function of medullary cells during urinary concentration but also during the rapid onset of diuresis. When exposed to hypotonicity a rapid release of these osmolytes occurs to counteract the cell swelling. The release occurs probably via channel-like proteins. Goldstein et al. reported on their studies related to taurine release in skate erythrocytes. The release pathway has the properties of a size-limited channel, the osmolyte must be less than 6.3 Å in mean molecular diameter. A positive charge (such as in choline) prevents even smaller molecules to pass. The channel selectivity appears to be based on size selection, taurine (5.5 Å) and betaine (5.8 Å) are transported faster than myo-inositol (6.2 Å). Inhibitor studies using three categories of inhibitors - (a) for anion exchangers (DIDS and PLP), (b) for Cl channels (NPPB, quinine, and MK-447A), and (c) for long chain fatty acids (saturated and unsaturated) further support the idea that the three chemical classes of organic osmolytes are released via the same swelling-activated channel. The anion exchanger band 3 seems to be closely linked to the release process. Studies on DIDS binding and the concentration of dimeric and tetrameric forms in swollen skate erythrocytes suggest that the oligomeric state of the anion exchanger might be directly related to changes in taurine transport. It remains to be determined whether band 3 acts as a channel or transmits the signal of cell swelling to a closely associated osmolyte channel of hitherto undetermined nature. Jackson and Strange have identified by patch clamp measurements swelling-activated anion channels in glioma cells and skate hepatocytes that also transport organic osmolytes, in particular taurine. This anion channel, termed VSOAC (volume-sensitive organic osmolyte/anion channel) is outwardly rectifying, has a unitary conductance of about 50 pS and a novel mechanism of activation. Single channels are switching one at a time from a completely OFF state into an ON state with very high open probability. A similar channel has also been found recently in rat papillary collecting duct cells, as reported by Kinne et al.. The same group demonstrated that at least in IMCD cells the release of each osmolyte is activated by different signal transduction pathways. Sorbitol release is triggered by a transient increase in intracellular calcium which involves arachidonic acid-dependent release from intracellular stores followed by the activation of calcium channels in the plasma membrane. There are further differences with regard to the involvement of G-proteins and the osmotic threshold for activation. In addition, the osmolyte release systems are distributed differently in the apical and basal plasma membrane. Sorbitol and betaine leave the cells exclusively through the basal-lateral membrane, whereas GPC, taurine, and myo-inositol use luminal and contraluminal transporters. The latter two seem to share at least one transport system. Thus, in contrast to the common signal for the accumulation under hypertonic conditions, the hypotonic release of each organic osmolyte seems to involve very special mechanisms of activation and specific transport systems.

Comparative Physiology of the Kidney*

Homer Smith made many pioneering contributions to the comparative physiology of the kidney and its evolution [H.W. Smith, in: Lecture of the Kidney, Univ. of Kansas, 1943; H.W. Smith, From Fish to Philosopher. Little, Brown, Boston, 1953]. One area of particular interest to him was the function of the aglomerular kidneys that can be found in about 30 species of mostly marine fish. The mechanism by which aglomerular fish form urine is, however, largely unknown to this date. The contribution by Beyenbach and Liu dealt with this mechanism. For the fluid secretion in the proximal tubule they propose that magnesium is actively secreted into the tubule lumen from which it cannot diffuse back into the blood. This transport causes the passive transepithelial secretion of diffusible sodium and chloride via the paracellular pathway. Water follows by osmosis. Since there is flow out of the distal end of the tubule a 'dynamic Donnan system' is maintained, driven by the active transport of magnesium. A mathematical model of tubular electrolyte and fluid secretion confirms this assumption. This model can also be employed to explain fluid secretion in proximal tubules of glomerular sea water fish, and thus describes a common, basic physicochemical mechanism for fluid transport in aglomerular and glomerular fish.

The magnesium to chloride ratio in the elasmobranch urine led Homer W. Smith in a paper published in 1931 [H.W. Smith, Am. J. Physiol. 98, 296-310, 1931] to conclude that the kidneys of elasmobranchs are not the only means to excrete salt. He stated "Since considerably more chloride is absorbed than magnesium (from the intestine) this fact shows that chloride has been excreted by an extrarenal route; and the excretion of chloride from a level of 270 mM per liter in the blood to a level of 500 mM per liter in sea water represents a process of osmotic work ..." This site of excretion has been identified to be the chloride cell of the gills and - where present - the rectal gland, that is analogous to the nasal 'salt' gland of the marine bird and reptiles. The rectal gland of elasmobranchs, in particular of the dogfish Squalus acanthias, has - as summarized by Silva et al. - provided the opportunity to elucidate various steps in the transepithelial transport of chloride and their application to many different epithelia, from amphibian cornea to the mammalian kidney. The uniformity of its cell population, the abundance of transport proteins and the variety of investigative pre-

---

* Indicates a title of a chapter in The Kidney: Structure and Function in Health and Disease

parations that can be utilized, provide a unique opportunity to examine the nature of chloride transport. The basic mechanism of secondary active chloride transport has been elucidated first in this gland and the molecular entities involved in the different steps have been identified. The regulation of chloride secretion is currently of major interest to the investigators. Silva et al. summarized the stimulatory and inhibitory factors. Volume expansion stimulates secretion in the intact animal. This stimulation is probably mediated by atrial natriuretic peptide (ANP) released from the atria and ventricle of shark heart. ANP, in turn, causes the release of VIP (vasoactive intestinal peptide) from nerve fibers in the parenchyma of the rectal gland. VIP activates the adenylyl cyclase and raises the cAMP level in the gland. This second messenger then enhances chloride secretion by increasing the activity of the luminally located CFTR-like chloride conductance and the basal-lateral Na-K-2Cl cotransporter.

Other related peptides such as the C-type natriuretic peptide (CNP) act directly on the rectal gland and bind to guanylyl cyclase-linked receptors, that have been cloned recently.

Neurotransmitters other than VIP mostly inhibit chloride secretion. This holds for somatostatin which exerts its action on adenylyl cyclase and on a site distal to the generation of cAMP. Bombesin causes the release of somatostatin from the nerves in the gland and is also inhibitory. Neuropeptide Y also reduces chloride secretion - probably distal to the site of cAMP generation. Thus a very intricate and complex pattern of regulation exists where stimulators and inhibitors coexist in nerves within the gland.

Another potent regulator of rectal gland function is adenosine. The cellular and molecular mechanisms of its action were discussed by Forrest. Rectal gland cells have a high density of both inhibitory A1-type receptors and stimulatory A2-type receptors. The former have a much higher affinity to adenosine than the latter. Therefore, adenosine inhibits at concentrations between 10 nM and 1 μM and stimulates chloride secretion at concentrations between 10 and 100 μM. This inhibition is probably an endogenous feedback regulation of chloride transport in the rectal gland. As chloride secretion increases adenosine concentration within the cells rises and adenosine is released by the cell into the interstitial space. There it binds to the high affinity A1 receptors and, in turn, inhibits transport. Thus adenosine functions in the rectal gland as an inhibitory autacoid to link energy demand to energy availability.

The other major salt excreting organ of the marine fish are the gills. The regulation of their transport activity during adaptation to different salinities was discussed by Zadunaisky using killifish (Fundulus heteroclitus) as example. This euryhaline fish can be transferred directly from fresh water to sea water and it survives and adapts to the situation. The signal received by the chloride cells in the gills to secrete more salt is the sudden increase in plasma osmolality produced by the transition from fresh water to sea water. Changes as small as 12.5 mOsm produce a 25% increase in chloride secretion in isolated preparations. This enables euryhaline fish to move back and forth from brackish water to sea water and regulate precisely their internal milieu.

Another particularly memorable event was the evening lecture at the MDIBL. During a glorious sunset a lobster and clam bake was served, followed by the presentation of Stephen Jay Gould entitled "New Insights from Comparative Biology". After briefly discussing and commenting upon Homer W. Smith's view on evolution he used this opportunity to elegantly and eloquently challenge the commonly held belief that "life history is an at least broadly predictable process of gradually advancing complexity through time" and to point to the important paleontological features that oppose this assumption. They are "the constancy of modal complexity throughout life's history, the concentration of major events in short bursts interspersed with long periods of relative stability; and the role of external impositions, primarily mass extinctions, in disrupting patterns of 'normal' times". According to Gould "these three features combined with the more general themes of chaos and contingency require a new framework of conceptualizing and drawing life's history" and thus he proposed a new iconography of evolution [S.J. Gould, Scientific American 271, 63-69, 1994]. This new iconography postulates that "the maximal diversity in anatomical forms - not in number of species - is reached very early in life's multicellular history". All major stages in organizing animal life's multicellular architecture occurred in a short period, beginning less than 600 million years ago and ending by about 530 million years ago - and the steps within this sequence are also assumed to be discontinuous and episodic. "This Cambrian explosion represents an initial filling of the 'ecological barrel' of niches for multicellular organisms. Later times feature extinction of most of these initial experiments and an enormous success within the surviving lines. This success is measured in the proliferation of species but not in the development of new anatomies. Each of these early experiments received little more than the equivalent of a ticket in the largest lottery ever played out on our planet - and each surviving lineage, including our own, inhabits the earth today more by luck of the draw than by any predictable struggle for existence".

Homer W. Smith drew quite a similar conclusion when he stated "There are those who say the human kidney was created" (i.e. did not develop) "... to keep our internal environment in an ideal balanced state. I would deny this - it owes the architecture, not to the design or foresight of any plan, but the geologic revolutions of 6000.000.000 years" [S.J. Farber, ibid].

REPORT TITLES

IMMUNOLOCALIZATION OF THE KINESIN RELATED PROTEIN (KRP$_{85/95}$) IN THE MIDPIECE AND FLAGELLUM OF SEA URCHIN (<u>STRONGYLOCENTROTUS DROEBACHIENSIS</u>) AND SAND DOLLAR (<u>ECHINARACHNIUS PARMA</u>) SPERM.

John H. Henson, Colleen D. Roesener, Stephanie Capuano, and Robert J. Mendola
Department of Biology, Dickinson College, Carlisle, PA 17013

Kinesin and its relatives comprise a large superfamily of homologous, ATP-dependent microtubule motor proteins which are thought to play important roles in microtubule-mediated vesicular/organellar transport and/or in mitotic spindle establishment and dynamics (for recent reviews see Skoufias and Scholey, <u>Curr. Op. Cell Biol.</u> 5: 95-104, 1993; Saunders, <u>Trends Cell Biol.</u> 3: 432-437, 1993). The best characterized kinesin-like protein in sea urchins is KRP$_{(85/95)}$, a plus end directed, heterotrimeric, kinesin related microtubule motor protein recently purified from sea urchin eggs (Cole et al., <u>J. Cell Sci.</u> 101: 291-301, 1992; Cole et al., <u>Nature</u> 366: 268-270, 1993). The KRP$_{(85/95)}$ complex consists of an uncharacterized 115 kDa subunit, and the 85 kDa and 95 kDa motor subunits. One or both of these latter subunits show sequence homology with the kinesin-like proteins encoded by the mouse KIF3a gene, the <u>Drosophila</u> KLP4 gene and the <u>Chlamydomonas</u> FLA10 gene. Using subunit-specific anti-KRP$_{(85/95)}$ monoclonal antibodies provided by our collaborators Drs. Jonathan Scholey and Douglas Cole (University of California at Davis), we have determined that KRP($_{85/95)}$ is present on membranous vesicle-like structures in the mitotic spindle of sea urchin embryos (Henson et al., <u>Dev. Biol.</u> 171:182-194, 1995), suggesting that KRP$_{(85/95)}$ may play a role in vesicle transport during mitosis.

Interestingly, recent studies have suggested that kinesin-like proteins may be important within flagella, a highly ordered microtubule-based structure previously thought of as the exclusive domain of the minus-end directed microtubule motor flagellar dynein. The <u>Chlamydomonas</u> FLA10 gene product is a kinesin-like protein, with homology to KRP$_{(85/95)}$, which appears to be involved in flagellar assembly and maintenance (Walther et al., <u>J. Cell Biol.</u> 126: 175-188, 1994). In addition the <u>Chlamydomonas</u> kinesin-like protein Klp1 localizes to one of the central pair microtubules in the axoneme of the flagellum (Bernstein et al., <u>J. Cell Biol.</u> 125: 1313-1326, 1994). The results of these recent studies prompted us to examine the expression and localization of KRP$_{(85/95)}$ in sea urchin and sand dollar sperm.

Sperm were shed from locally collected <u>S. droebachiensis</u> sea urchins and <u>E. parma</u> sand dollars by intracoelomic injection of 0.5 M KCl. For immunoblotting, pellets of whole sperm cells were either directly lysed in SDS sample buffer, or the cells were dounce and ultrasonically homogenized and then diluted in sample buffer. Some samples were treated with DNase in order to break up DNA and lower the viscosity of the samples. In addition, gel samples were generated of sperm tails obtained by forcing sperm suspensions through a 25 gage needle several times prior to differential centrifugation. All gel samples were subjected to SDS-polyacrylamide gel electrophoresis and then transferred onto nitrocellulose. The nitrocellulose filters were then probed with either a mixture of anti-KRP$_{(85/95)}$ monoclonal antibodies (K2.2, K2.3, K2.5; see Cole et al., 1993. loc cit.) or with an affinity purified anti-115 kDa polyclonal antiserum, followed by treatment with the appropriate alkaline phosphatase conjugated secondary antibodies. For immunofluorescent staining, sperm cells adhered to poly-lysine coated coverslips were fixed with -20º C methanol plus 40 mM EGTA, incubated with either individual or mixes of anti-KRP$_{(85/95)}$ antibodies (at a concentration of 100 µg/ml each) or the 115 kDa antiserum, and then incubated in fluorophore conjugated secondary antibodies. In double labeling experiments sperm microtubules were labeled using either an anti-sea urchin tubulin polyclonal antiserum or an anti-alpha tubulin monoclonal. Other labeling experiments utilized a commercial anti-flagellar dynein intermediate chain monoclonal antibody (Sigma Chemical Co.), an anti-centrosome monoclonal antibody raised against <u>Drosophila</u> intermediate filaments (graciously provided by Dr. Calvin Simerly, University of Wisconsin at Madison; see Schatten et al., <u>Proc. Natl. Acad. Sci. USA</u> 84: 8488-8492, 1987), the Hoechst dye 33258 to label nuclei, and the mitochondria-specific vital dye rhodamine 123. Fluorescent specimens were viewed on a Nikon Optiphot II epifluorescence microscope using a 60X (NA 1.4) planapo phase contrast objective lens, and 35 mm photographs were taken using Kodak TriX ASA 400 film.

Immunoblotting results indicated that sperm samples probed with the anti-KRP$_{(85/95)}$ antibodies contained immunoreactive proteins which comigrated with the 85 kDa and the 115 kDa subunits of control sea urchin egg samples (see Figure 1). These immunoreactive species were present in samples of whole sperm and isolated sperm tails.

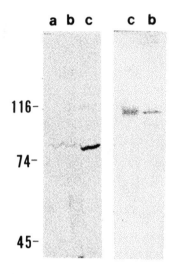

<u>Figure 1:</u> Immunblot of anti-KRP$_{(85/95)}$ monoclonals K2.2-2.5 (left panel) and the anti-115 kDa polyclonal (right panel) against samples from sea urchin sperm (lane a), sand dollar sperm (lane b) and sea urchin eggs (lane c). Molecular masses of standards are given in kDa on the left.

Immunofluorescent localization of KRP$_{(85/95)}$ in sperm revealed faint labeling of the flagellum and more intense staining of the midpiece (Figure 2). The midpiece is defined as the portion of the sperm lying between the nucleus and the flagellum which contains the centrosome, many mitochondria, and the majority of the sperm's cytoplasm. The midpiece-specific nature of the KRP$_{(85/95)}$ staining pattern was confirmed in cells triple labeled for KRP$_{(85/95)}$, microtubules and nuclei. In addition, the KRP$_{(85/95)}$ midpiece staining pattern localized to the same region that was labeled by either the centrosome-specific antibody, or the mitochondria-specific dye rhodamine 123.

The KRP$_{(85/95)}$ labeling of the midpiece and flagella often appeared punctate suggestive of an association with cytoplasmic vesicles. This pattern of punctate labeling was different from the more continuous, linear pattern seen in the flagella of sperm labeled with anti-flagellar dynein intermediate chain. An association between KRP$_{(85/95)}$ and membranous vesicles is also suggested by the loss of staining intensity seen in sperm extracted prefixation with Triton X-100 detergent under microtubule stabilizing conditions (see Wright et al., <u>J. Cell Biol.</u> 113: 817-833, 1991). Flagellar dynein labeling of sperm tails was not affected by prefixation extraction.

The results of the present study suggest that the kinesin-like heterotrimeric protein KRP$_{(85/95)}$ is associated with membranous structures in the midpiece and flagella of echinoderm sperm. It is interesting to note that a very recent meeting abstract (Kozminski et al., <u>Mol. Biol. Cell</u> 6: 253a, 1995) reports genetic and immunolocalization results which indicate that the KRP$_{(85/95)}$ homologue in <u>Chlamydomonas</u>, FLA10, is associated with the intraflagellar transport of granule-like particles. These results are consistent with our localization findings reported here. We are currently attempting to purify KRP$_{(85/95)}$ from sea urchin sperm in order to begin a biochemical characterization of the protein.

(The authors gratefully acknowledge the assistance of Dr. Ray Rappaport in showing us how to obtain sperm from sand dollars. Supported by NIH grant GM47693-01, a NSF Young Investigator Award (MCB-9257856), and NSF ILI grant USE-9250861 to JHH, and a Dana Internship Award from Dickinson College to RJM.)

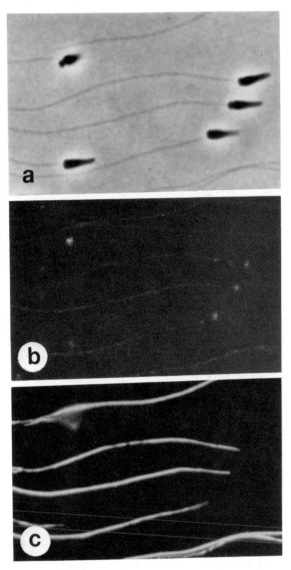

<u>Figure 2:</u>   Immunofluorescent labeling of the KRP$_{(85/95)}$ 115 kDa subunit (panel b) and microtubules (panel c) in sand dollar sperm. Panel a shows a phase contrast image of the sperm. Note that the 115 kDa antiserum labels both the midpiece and the flagella of the sperm in a punctate manner. Magnification = 1,200X

WIDTH OF THE CONTRACTILE REGION IN SAND DOLLAR (<u>ECHINARACHNIUS PARMA</u>) EGG CLEAVAGE FURROWS

R. Rappaport
The Mount Desert Island Biological Laboratory
Salsbury Cove, ME 04609

The tip of the active furrow in animal cells is underlain by a ring of circumferentially oriented microfilamentous actin approximately 10 µm wide and 0.1 - 0.2 µm thick (Schroeder, T. E., J. Cell Biol. <u>53</u>, 419-434, 1972). This ultrastructurally distinctive region is generally assumed to be the force producing mechanism that divides the cell by contraction and that its characteristic structure is necessary to produce the force that deforms the cell. The purpose of this investigation was to determine the width of the equatorial cortex region that is capable of autonomous equatorial constriction and compare it with the width of the ultrastructurally demonstrable contractile ring.

Gametes were obtained by KCl injection and the hyaline layer was removed from fertilized eggs by glycine treatment or multiple rinses with calcium-free artificial sea water. When cleavage began, apposed stout needles were inserted through the polar regions and oriented parallel to and in the same plane as the mitotic axis (figure 1).

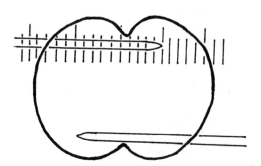

Figure 1. Diagrammatic representation of the relation between the sand dollar egg, the needles and the ocular micrometer. Not to scale.

When the underside of the cortex at the furrow tip contacted a needle its progress toward the mitotic axis was locally halted. The needles did not move. As contraction of the unblocked furrow regions progressed, its shape changed from a circle to an ellipse to parallel strips stretched between the needles. Concurrently the width of the equatorial cortex in contact with the needle widened. The mean of the widths of the flattened regions of 22 eggs was 25.6 ± 4.4 µm. The extremes were 15 and 31 µm.

The flattening of the furrow tip resulted from the resistance of the needles. Had the capacity for autonomous equatorial constriction been restricted to the region where the microfilaments are circumferentially oriented, the flattened region would have been about 10 µm wide. The greater width strongly suggests that force production sufficient for deformation is possible in cortical regions that lack oriented microfilaments so that a portion of the furrow wall, as well as the tip, are actively contractile. Since calculations of contractile tension exerted per unit cross sectional area of the furrow are based upon the ultrastructural dimensions of the contractile ring, its magnitude may have been overestimated about two fold.

Supported by National Science Foundation Grant DCB - 9416654

MECHANISMS OF SILVER ION (Ag⁺) TOXICITY IN FERTILIZED EGGS OF ILYANASSA OBSOLETA

MECHANISMS OF SILVER ION (Ag^+) TOXICITY IN FERTILIZED EGGS
OF ILYANASSA OBSOLETA

G.W.Conrad[1], M.J.Janasek[1], N.M.Martinez[2], and A.H.Conrad[1]
[1]Division of Biology, Kansas State University, Manhattan, KS
[2]Department of Biology, Georgia Southern University, Statesboro, GA

We have demonstrated previously that microtubule distribution in fertilized eggs of the common marine mudsnail, Ilyanassa obsoleta Stimpson (=Nassarius obsoletus Say) is very sensitive to the presence of silver ions (Ag^+) in the sea water (A. Conrad, et al. Cell Motil. & Cytoskel. 27: 117-132 (1994)): a narrow range of Ag^+ concentrations ($5-7 \times 10^{-11}M$) causes a marked increase in the numbers of microtubules in a normally transient cytoplasmic neck (polar lobe constriction), followed by great elongation of the neck and its eventual severing, an abnormal developmental event. This response largely mimics the cellular response to the reference standard microtubule stabilizing agents, taxol and hexylene glycol (A. Conrad et al. J. Exp. Zool. 262: 154-165 (1992) and J. Exp. Zool. 269: 188-204 (1994)). Heavy metal ions, such as Ag^+, are thought to interact with proteins via a cage formed from three sulfhydryl groups. We therefore asked if any other metal ion could duplicate the Ag^+ response and whether it could be mimicked by agents that could form cross-links between sulfhydryl groups, in the absence of Ag^+.

Effects of other metal ions: In a narrow range of concentrations (0.75-1.5 μM), we observed that Cu^{2+} causes 13- 37% of Ilyanassa fertilized eggs to form very long polar lobe necks resembling those formed in response to Ag^+. This indicates that Ag^+ is not the only heavy metal ion to cause this effect.

Effects of cross-linking reagents: Four homobifunctional sulfhydryl cross-linking reagents were assessed for their ability to elicit Ag^+-like cellular deformation of fertilized Ilyanassa eggs (very long, thin polar lobe necks; a shape not seen during normal development): N,N'-p-phenylene dimaleimide (p-PDM), N,N'-bis(3-maleimidopropionyl)- 2-hydroxyl-1,3-propanediamine (N,N'-bis), Bismaleimidohexane (BMH), and 1,4-Di-[3'-(2'pyridyldithio)propionamido]butane (DPDPB). When applied to cells in sea water, p-PDM, N,N'-bis, and BMH either caused no cells to form abnormally elongated polar lobe necks, or caused very few to form (N,N'-bis; <5% of cells). In contrast, exposure of cells to DPDPB caused as many as 47% of cells to assume this unusual shape. The range, 75-250 μM, causes an average of 6 % or more cells to form long necks, with the optimum concentration of 125 μM causing an average of 11% with long necks. If eggs are pretreated with a reducing agent, dithiothreitol, followed by DPDPB, the percentage of responding cells increases to 24%, whereas if the pretreatment is with an oxidizing agent, H_2O_2, the percentage of responding cells is 19 % (continuous treatment with dithiothreitol gives only 3% responding cells, whereas continuous exposure to H_2O_2 gives 4% responding). We conclude that Ag^+-like morphological effects can be produced by treatment with a homobifunctional sulfhydryl reactive cross-linking reagent. (Support: NASA NAGW-4491, NASA-NSCORT NAGW-2328, & NSF REU 9322221)

MAITOTOXIN ACTIVATES A NON-SELECTIVE CATION CHANNEL IN CARDIAC MYOCYTES OF <u>RATTUS NORVEGICUS</u> AND <u>SQUALUS ACANTHIAS</u>

James Maylie[1], Daggett Harvey[2], Francis M. Van Dolah[3], and Martin Morad[4]
[1]Department of Obstetrics and Gynecology, OHSU, Portland, OR 97201
[2]Wesleyan College, Middletown, CT 06459
[3]Charleston Laboratory of the U.S. National Marine Fisheries Services, Charleston, SC 29412
[4]Department of Pharmacology, Georgetown University, Washington DC 20007

Maitotoxin (MTX) isolated from the marine dinoflagellate <u>Gambierdiscus toxicus</u> is a water-soluble polyether that may be responsible for ciguatera seafood poisoning (Yokoyama et al., J. Biochem, 104:184, 1988). The toxic effects of MTX presumably result from calcium influx that evoke transmitter release, muscle contraction, and phosphoinositide breakdown in a variety of cell types (Gusovsky & Daly, Biochem. Pharm. 39:1633, 1990). Initial studies suggest that the MTX induced calcium influx is mediated by dihydropyridine (DHP) sensitive and insensitive pathways with the DHP insensitive pathway being blocked by cadmium and SK&F 96365 (a receptor and voltage activated Ca^{2+} channel antagonist) in some cell types (Soergel, et al, Mol. Pharm., 41:487, 1992). This lead to the idea of a ubiquitous MTX-sensitive calcium channel. However, more recent studies suggest that MTX activates non-selective cation, sodium or chloride channels in epithelial or ß-cell lines (Dietl & Völkl, Mol. Pharm., 45:300, 1994, Worley, et al., J. Bio. Chem., 269:32055, 1994). Thus the mechanism by which MTX stimulates Ca^{2+} entry is still poorly understood. In this study we show that MTX does not affect T- or L-type Ca^{2+} currents in dogfish cardiac myocytes. Rather, MTX activates a non-selective cation channel with a reversal potential near -6 mV under physiological conditions in both dogfish and rat cardiac myocytes.

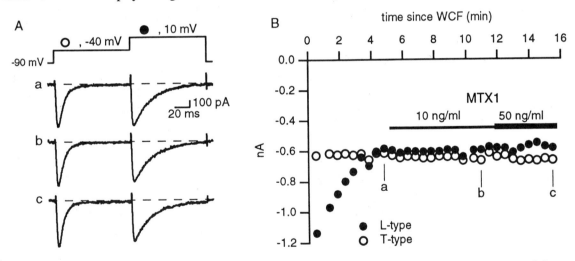

Figure 1. Lack of effect of MTX on Ca^{2+} currents in shark myocyte. A. T- and L-type I_{Ca} measured with a dual pulse protocol. The potential was stepped from a holding potential of -90 mV to -40 mV for 100 ms then to 10 mV for 100 ms (upper trace). Current traces a-c recorded in control solution, 10 ng/ml MTX , and 50 ng/ml MTX, respectively. Currents were leak corrected with a P/4 procedure. B. The peak inward current at -40 mV (closed symbols) and 10 mV (open symbols) is plotted versus time since WCF. The bars indicate exposure time to 10 and 50 ng/ml MTX, respectively. Letters correspond to traces in panel A.

The effect of MTX on Ca^{2+} currents (I_{Ca}) was tested in dogfish cardiac myocytes since this species expresses both T- and L-type calcium channels with approximately equal current density. Shark ventricular myocytes were isolated using established procedures and studied with the whole cell patch clamp configuration. To isolate Ca^{2+} currents from Na^+ and K^+ currents, the patch pipette contained (mM): CsCl 240, $MgCl_2$ 1, Urea 300, EGTA 20, HEPES 20, TMAO 70,

MgATP 5, pH 7.2. Cells were initially superfused with shark Ringer (mM): NaCl 270, KCl 4, MgCl$_2$ 3, KH$_2$PO$_4$ 0.5, Na$_2$SO$_4$ 0.5, Urea 350, HEPES 10, glucose 10, CaCl$_2$ 3, pH 7.2. Following formation of whole cell recording configuration (WCF) the external solution was switched to a Na$^+$ and K$^+$ free solution: TEA-Cl 275, CaCl$_2$ 5, MgCl$_2$ 5, HEPES 10, Urea 350, pH 7.2. Ionic currents were measured with an EPC-9 (HEKA). Figure 1 shows that MTX did not affect T- and L-type I$_{Ca}$. Ca^{2+} currents were measured with a double-step procedure in which the membrane potential was first stepped from a holding potential of -90 to -40 mV and then to 10 mV to sequentially activate T- and L-type I$_{Ca}$, respectively. The T-type I$_{Ca}$, activating at -40 mV, inactivated rapidly during the pulse and did not interfere with activation of the L-type channel during the second step to 10 mV. The plot of peak inward T- and L-type I$_{Ca}$ as a function of time following WCF shows that the L-type I$_{Ca}$ decreased during the first 4 min following WCF as a result of run-down (Fig. 1B). Addition of 10 or 50 ng/ml of MTX, diluted from a stock solution of 500 µg/ml in methanol, to the bath solution had no effect on T- or L-type I$_{Ca}$ (Fig 1A, B). Similar results were observed in 7 out of 8 cells. However, in one cell 10 ng/ml of MTX decreased the T-type I$_{Ca}$ by 17% and increased the L-type I$_{Ca}$ 61%. In this experiment, MTX was diluted from a stock solution of 10 µg/ml in methanol. It was not determined whether the increase in L-type I$_{Ca}$ in this cell was the result of 0.1% methanol which increases L-type I$_{Ca}$.

These results show that MTX does not affect T- or L-type Ca^{2+} channels in dogfish heart. To determine whether the lack of effect of MTX on cardiac Ca^{2+} channels is species specific, similar experiments were carried out on rat cardiac myocytes which express only L-type Ca^{2+} currents. Rat cardiac myocytes were isolated using established procedures and perfused with Tyrode solution containing 0.2 mM Ba^{2+} to block inward K$^+$ currents. The composition of the Tyrode solution was (mM): NaCl 140, KCl 5, MgCl$_2$ 1, CaCl$_2$ 2, HEPES 10, glucose 10, pH 7.4. The patch pipette contained (mM): CsCl 30, CsAsp 110, EGTA 20, HEPES 20, MgATP 5, pH to 7.2 with CsOH. Addition of MTX following whole cell formation appeared to have no effect on I$_{Ca}$ in that no increase over the apparent run-down I$_{Ca}$ was observed. Rather, the input conductance increased during perfusion with MTX. Figure 2A shows a family of current traces in control and 21 min after perfusion with 10 ng/ml of MTX. The plot of final current versus voltage shows that MTX activated a current with a linear current-voltage relation that reversed near -6 mV (panel B). That both Na$^+$ and Ca^{2+} currents were still observed suggest that a loss of patch pipette seal cannot account for the effect of MTX.

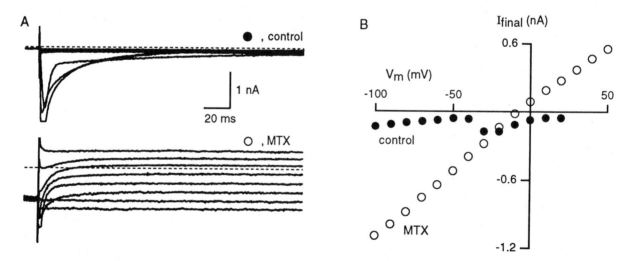

Figure 2. MTX activates a channel with a linear conductance in rat cardiac myocyte. A. Family of current traces in control (upper traces) and after 21 min perfusion with 10 ng/ml MTX (lower traces). Currents were recorded from a holding potential of -80 mV to test potentials between -100 and 20 mV (control) and -100 and 40 mV (MTX) in 20 mV increments. Dash line represents zero current. B. Plot of final current measured at the end of the test pulse versus test pulse potential.

The effect of MTX in rat was dose dependent with 10 ng/ml giving a greater increase in conductance than 1 ng/ml (Figure 3). The membrane potential was initially depolarized to 60 mV to inactivate Na^+ and Ca^{2+} currents and subsequently slowly repolarized to -120 mV over 1 s yielding a quasi steady state current-voltage relation. The control ramp shows residual Ca^{2+} currents that were absent in the MTX records as a result of run-down. During perfusion with MTX, the conductance increased 0.3 and 2.7 nS/min with 1 and 10 ng/ml MTX, respectively. MTX, 10 ng/ml, produced an increase in the steady state conductance within 1 min of its application and the average rate of increase in the conductance was 3.9 ± 0.8 nS/min (mean±SEM, n=6). The effect of MTX was reversible only with prolonged washout (>20 min). Steady state activation and a complete dose response of MTX was not determined.

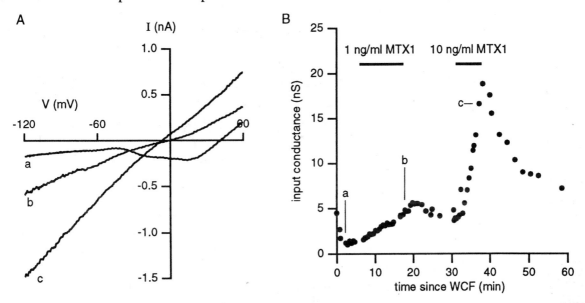

Figure 3. Ramp protocol to measure MTX activated current in rat cardiac myocyte. A. Quasi steady state current-voltage relation obtained from a voltage ramp of 60 to -120 mV in 1 s. Ramps a-c recorded in control solution, 1 ng/ml MTX, or 10 ng/ml MTX, respectively. B. Plot of input conductance versus time since WCF. Input conductance was measured with a -5 mV step from a holding potential of -50 mV. Letters correspond to ramps in panel A.

Ionic substitutions experiments were performed to determine the nature of the charge carrier of the MTX activated current in rat ventricular myocytes. To test whether Cl^- is a charge carrier, the external $[Cl^-]$ was decreased from 153 to 11.4 mM (Cl^- replaced with methansulfonate); the Cl^- reversal potential would shift from -41 to 25 mV. The reversal potential of the MTX activated current was -6 mV in both normal and low Cl^-. Thus the MTX activated current is not carried by Cl^- ions in cardiac myocytes.

Rather, the reversal potential of ~-6 mV is more consistent with MTX activating a non-selective cation channel. The ion selectivity of the MTX activated current was determined under biionic conditions in which the bath solution was replaced with an external solution containing (mM): mannitol 280, HPES 1, $BaCl$ 0.2, pH 7.4 with NaOH plus either NaCl 20, CsCl 20, or KCl 20. Following activation of the MTX sensitive current in normal Tyrode plus 0.2 BaCl, the external solution was switched to the 20 mM Cs-mannitol solution (Fig 4A). The reversal potential shifted from -9 in Tyrode to -45 mV in 20 Cs-mannitol; the predicted E_{Cs} for this solution is -56 mV and the measured reversal potential was not corrected for shifts in the junction potential. The reversal potential in the 20 Na- and 20 K-mannitol solution was -42.0 and -45 mV, respectively. The reduction in current in the Na- and K-mannitol solution is in part due to wash of MTX which was not included in the mannitol solutions. Fig 4B shows that the MTX activated current in the

9

Na-mannitol solution which does not contain Ca^{2+} is voltage and time independent. The permeability ratios calculated from the equation, $\Delta E_{rev} = 58.6\log(P_x[x]_o/P_{Na}[Na]_o)$ gave a value of 0.9 for both P_K/P_{Na} and P_{Cs}/P_{Na}. A second cell gave a permeability ratio of 0.94 for P_K/P_{Na} and P_{Cs}/P_{Na}. The permeability ratio for tetraethylammonium (TEA) was indirectly determined by replacing the NaCl and KCl in the normal Tyrode solution with TEA-Cl (Fig. 4C). The inward current was significantly reduced, shifting the reversal potential from -10 to -44 mV. The estimated P_{TEA}/P_{Na} was 0.25.

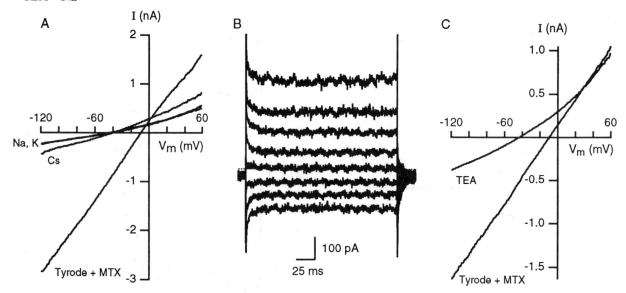

Figure 4. Cation selectivity of MTX activated current in rat cardiac myocyte. A. Current-voltage relations obtained from voltage ramps from 60 to -120 mV in Tyrode plus 10 ng/ml MTX, after 3 min perfusion with Cs-mannitol (Cs), 5 min with Na-mannitol (Na), and 2 min with K-mannitol (K). The IV for Na- and K-mannitol are overlapping. B. Family of current traces in Na-mannitol solution evoked from a holding potential of -50 mV to test potentials from -100 to 40 mV in 20 mV increments. C. Current-voltage relations obtained from voltage ramps from 60 to -120 mV in Tyrode plus 10 ng/ml MTX and after 2 min perfusion with TEA-Cl solution (TEA).

These results show that MTX activates a non-selective cation channel in rat ventricular myocytes. A similar non-selective cation current is activated by MTX in shark ventricular myocytes (data not shown). It is thus proposed that MTX stimulates Ca^{2+} entry into cardiac myocytes secondarily to activation of a non-selective cation channel. Activation of the non-selective cation channel depolarizes the membrane potential towards its reversal potential of ~-6 mV thus activating Ca^{2+} channels and reducing Na^+-dependent Ca^{2+} efflux from the cell by the electrogenic Na^+-Ca^{2+} exchanger. These effects will result in an increase in intracellular Ca^{2+} and muscle contraction. Consistent with this hypothesis is the observation that non-patched cells shorten when exposed to MTX.

This research was sponsored by a fellowship from the NIEHS Toxicology Center to JM and by a grant from AHA Maine to MM.

AN EXOCYTOTIC RELEASE PATHWAY TRIGGERED BY Ca^{2+} INFLUX VIA Ca^{2+} CHANNELS IN CHROMAFFIN CELLS FROM RATTUS NORVEGICUS.

Jing Fan, Lars Cleemann and Martin Morad.
Department of Pharmacology, Georgetown University, Washington, DC. 20007.

Secretion of catecholamine in bovine chromaffin cells is thought to occur primarily when exocytosis of docked secretory vesicles is triggered by μ-domains of Ca^{2+} which develop to a concentration of several μM near the inner opening of Ca^{2+} channels (Augustine and Neher, J. Physiol. 450:247-271, 1992). It is also recognized that the global, as well as the local, intracellular concentration of Ca^{2+} may cause exocytosis or determine the size of the pool of release-ready, docked vesicles. Furthermore, it is thought that intracellular Ca^{2+} stores contribute to the control of secretion, that different types of Ca^{2+} channels (L,P,Q etc.) may not trigger secretion equally, and that both the internal stores and the diversity of Ca^{2+} channels vary from species to species.

In the present study we have examined the control of secretion in single rat chromaffin cells under stringent experimental conditions where the time course and magnitude of the secretory response was compared directly to both Ca^{2+} current and global intracellular Ca^{2+} concentration. For this purpose we used simultaneously the whole-cell voltage clamp technique to measure Ca^{2+} current (I_{Ca}), fluorescent dyes to detect intracellular Ca^{2+} transients (Ca_i-transients), and mono-filament carbon fiber electrodes to record the current resulting from oxidation of released catecholamines.

Primary cultures of rat chromaffin cells were prepared by enzymatic digestion followed by separation on a sucrose gradient (Fan, Cleemann, Lara, Gandia and Morad, Bull. MDIBL 34:12-13, 1995). The cells were suspended in Dulbecco's modified Eagle's medium supplemented with 5% fetal calf serum, 50 IU/ml penicillin and 50 μg/ml streptomycin and were plated onto glass cover slips which had been coated with 50 μg/ml collagen and then dried and baked. Experiments were performed at room temperature after culturing from 2 to 6 days. The standard external solution contained (in mM): 125 NaCl, 1 $MgCl_2$, 2 $CaCl_2$, 10 glucose and 10 HEPES at pH 7.4. K^+ was omitted from this solution in order to suppress Ca-activated K^+ current. The whole cell patch-clamp electrodes had a resistance of 4-5 MΩ and were filled with a dialyzing solution containing (in mM): 80 Cs-aspartate, 30 CsCl, 5 Mg-ATP, 0.05-0.2 Ca-indicator dye and 10 HEPES at pH 7.2. The Ca-indicator dyes (Calcium-green or Fluo-3) were excited with epi-illumination using 488 nm light from an argon laser and emitted light was recorded with photo-multiplier detector placed behind a 500 nm glass barrier filter. This method allows measurements of relative rather than absolute increases in cytosolic calcium. The carbon fiber electrodes used for the recording of exocytosis of single catecholamine containing vesicles were insulated with polyethylene tubing, were biased with 600 mV, and had an exposed 8 μm tip which was pressed gently against one side of the voltage-clamped cell (diameter 10-20 μm, capacitance 8-25 pF). The secretion was measured as the average oxidation current recorded typically during 3 to 10 voltage clamp pulses.

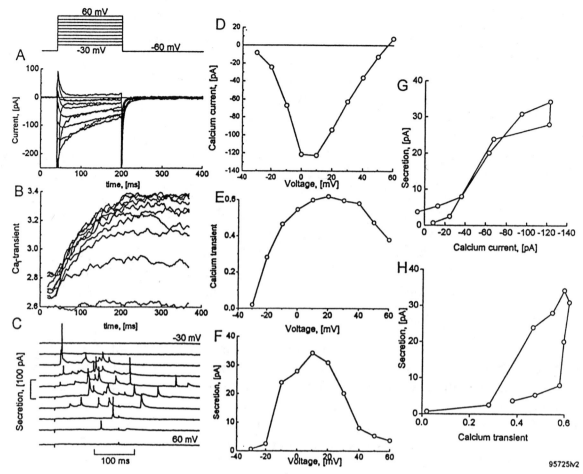

95725N2

<u>Figure 1.</u> Voltage dependence of I_{Ca} (panels A, D and G), Ca_i-transients (panels B, E, and H), and secretion (panels C, G, F and H). Original records are shown on the left, voltage relations in the middle and correlations on the right. The Ca^{2+} current (panel D) was measured 40 ms after depolarization. The Ca_i-transients (panel E) are uncalibrated and were measured as the change in fluorescence from the start to the finish of the clamp pulse. Secretion (panel F) is the average oxidation current calculated as the integral charge of the secretory spikes divided by the clamp duration (160 ms).

The voltage dependencies of I_{Ca}, the Ca_i-transient and the secretory response are illustrated in Figure 1. The left side of the figure shows original records from a series of 160 ms voltage clamp pulses where the membrane was clamped, in 10 mV increments, to potentials in the range from - 30 to +60 mV. I_{Ca} is seen as the slowly inactivating component which follows the initial capacitive current and Na^+ current (panel A). During the clamp pulse the Ca_i-transients rose, first rapidly and then more slowly (panel B). Notice that the decay of the Ca^{2+} transients is barely noticeable during the first 200 ms following repolarization. In fact, more than 5 sec was generally required for Ca^{2+} transients to decay 90%, and there was often, as in the present case, some carry-over from one depolarization to the next, even when they were separated by 30 sec. The recorded oxidation currents clearly show the current spikes associated with individual secretory vesicles (panel C). Unlike the Ca^{2+} transients, the secretion rapidly fell to zero when I_{Ca}

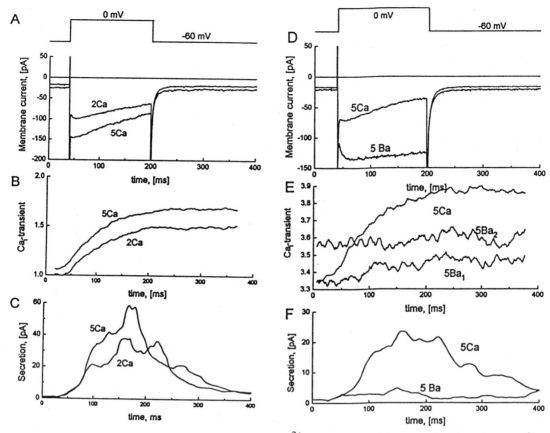

Figure 2. Effect of increased and blocked Ca^{2+} current on Ca_i-transient and secretion. From top to bottom the panels show the membrane current (panels A and D), the fluorescence signal (panels B and E) and the secretion (panels C and F). The left side of the figure shows the results of increasing the extracellular Ca^{2+} from 2 to 5 mM. The right side shows the effect of substitution of 5 mM Ba^{2+} for 5 mM Ca^{2+}. Panel E shows the average Ca_i-transient in the presence of Ca^{2+} and the two first recordings ($5Ba_1$ and $5Ba_2$) after the change to Ba^{2+} containing solution. The oxidation currents are averages from a number of depolarizations and over a 40 ms window.

was terminated by repolarization. The middle panels of Fig. 1 show the voltage dependencies of the three parameters (panels D,E and F). In all cases we found a bell shaped relationship with a maximum near 10 mV and smaller values at both higher and lower potentials. Some significant differences are revealed, however, when the three parameters are in turn compared (panels G and H). The secretion was proportional to the Ca^{2+} current (panel G), but showed significant non-linearity and hysteresis when compared to the Ca^{2+} transient (panel H). We conclude that both the time course and voltage-dependence of the secretory response followed the Ca^{2+} current much more closely than the Ca^{2+} transient.

A causal relationship between Ca^{2+} current and secretion was also found in experiments with double pulses and clamps of different duration. For instance, clamp pulses of 50, 250 and 450 ms duration , produced secretory responses of similar durations. This showed that the secretion lasts

13

as long as the Ca^{2+} current, and dispels the possibility that the termination of secretion (as in fig. 1C) might be due to depletion of a pool of release-ready vesicles.

It might be argued that saturation of the Ca^{2+} indicator dye could account for its broader bell shaped voltage-dependence (Fig. 1E). Against this idea, however, we found that the indicator dye registered a higher Ca^{2+} concentration than during previous depolarizations when the seal of the patch electrode was lost. Furthermore, the Ca^{2+} transients at 0 mV increased in amplitude when the Ca^{2+} current (Fig. 2, panel A) increased following elevation of the extracellular Ca^{2+} concentration from 2 to 5 mM (panel B). Notice that this intervention increased I_{Ca} and secretion proportionally.

The obligatory role of Ca^{2+} influx in the secretory response was demonstrated in experiments where secretion was completely blocked (Fig. 2, panel F) when extracellular Ba^{2+} was substituted for extracellular Ca^{2+} as charge carrier through the Ca^{2+} channel (Fig. 2, panel D). Panel E of Fig. 2 suggests that entry of Ba^{2+} into the cell caused an increase in the fluorescence of the indicator dye.

The present study of rat chromaffin cells support and extend results from bovine chromaffin cells. The extensive use of simultaneous measurements of I_{Ca}, Ca_i-transients and single secretory events is a novel feature which makes it possible to guard against a number of artifacts (rundown of I_{Ca} and Ca^{2+} overload) and yields highly significant results with high sensitivity and excellent time resolution. The different voltage-dependencies of I_{Ca} and Ca_i-transient is a new finding. The major result is that both magnitude and duration of secretion in rat chromaffin cells is controlled directly by the Ca^{2+} current and is influenced very little, if at all, by the global intracellular Ca^{2+} concentration. On the other hand, it is not known why the voltage dependence of Ca_i-transients differs from that of the Ca^{2+} current. A likely explanation is that the Ca_i-transients, in part, reflect other Ca^{2+} pathways such as release of Ca^{2+} from internal stores, or entry via a Na-Ca exchanger. Our results are consistent with the idea that secretion is controlled by μ-domains of Ca^{2+}

Supported by NIH RO1 16152 and the Maine Affiliate of the American Heart Association.

MICROMOLAR CONCENTRATIONS OF INORGANIC MERCURY ALTER MEMBRANE CONDUCTANCE OF <u>XENOPUS</u> OOCYTES

James A. Schafer, Jr.[1], Osak Omulepu[2],

Beatrice Chen[1], Sudhakar Cherukuri[1] and

David C. Dawson[1]

[1]Department of Physiology, University of Michigan
Medical School, Ann Arbor, MI 48109-0622 and

[2]Morehouse College, Atlanta, GA

In the course of experiments in which Xenopus oocytes were employed as an expression system for the study of the effects of inorganic mercury on cloned transporters, we investigated the effects of mercury on the background ionic permeability of the oocyte plasma membrane. Oocytes were removed from frogs and defolliculated as previously described (L.S. Smit, D.J. Wilkinson, M.K. Mansoura, F.S. Collins, and D.C. Dawson, Proc. Nat. Acad. Sci. 90:9963-9967, 1993). During experiments oocytes were perfused with a modified Barth's solution that contained (in mM): 98 NaCl, 2 KCl, 1.8 $CaCl_2$, 1 $MgCl_2$ and 5 Hepes (pH 7.4, 220 mOsm). A two electrode voltage clamp was used to monitor membrane conductance. Current-voltage relations for the oocyte membrane were obtained by means of a ramp command that varied the clamping potential from -120 mV to +60 mV over a two second interval. The holding potential was generally -60 mV. Currents and voltages were acquired by means of an IBM compatible computer using "pCLAMP" software (Axon Inst, Foster City, CA).

Figure 1 shows a representative I-V plot for an oocyte under control conditions and after exposure to 1 µM $HgCl_2$ for 9 min. Under control conditions the I-V plot for the oocyte exhibited a characteristic appearance. At negative membrane potentials the plot was linear with a reversal potential of about -30 mV. At positive potentials the curve was S-shaped due to the activation of an endogenous, calcium-activated, chloride-selective conductance (D.J. Wilkinson, M.K. Mansoura, P.Y. Watson, L.S. Smit, F.S. Collins, and D.C. Dawson, J. Gen. Physiol. 107:103-119, 1996). Exposure of the cell to 1 µM $HgCl_2$ in frog Ringer's, shifted the reversal potential toward more negative values and increased the slope conductance at the reversal potential. The effect of $HgCl_2$ was not reversed by washing the oocyte with $HgCl_2$-free frog Ringer's and was only partially reversed by the application of 100 µM dithiothreitol.

The Effect of 1µM $HgCl_2$ on Whole Cell Conductance

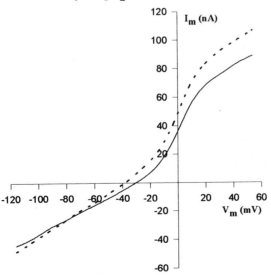

FIGURE 1. Current-voltage plots for <u>Xenopus</u> oocyte in the presence of frog Ringer's (solid line) and after exposure to 1µM $HgCl_2$ in frog Ringer's (dashed line). In the absence of $HgCl_2$, the membrane current reversed at -33.2mV and the slope conductance at the reversal potential was 0.591µS. Exposure to $HgCl_2$ shifted the reversal potential to -39.1mV and increased the slope conductance to 0.712µS.

These results indicate that micromolar quantities of $HgCl_2$ can significantly alter the ionic conductance of the oocyte plasma membrane. The increase in slope conductance and leftward shift in the reversal potential are compatible with the hypothesis that the target for $HgCl_2$ was a population of potassium-selective channels. The observed response could be accounted for by increases in either the single channel conductance or the open probability of such channels, or by some combination of these effects. These findings suggest that micromolar concentrations of $HgCl_2$ might compromise cellular processes like volume regulation by activating potassium channels. In addition, the effect of Hg^{2+} on endogenous oocyte potassium channels could complicate any evaluation of the metal's action on cloned channels expressed in oocytes (Supported by NIEHS E503829).

MOLECULAR CHARACTERIZATION OF MYOSIN AND THE SODIUM-PROTON ANTIPORTER IN ILYANASSA OBSOLETA

A.H. Conrad & G.W. Conrad
Division of Biology, Kansas State University, Manhattan, KS 66506

Ilyanassa obsoleta Stimpson, the common marine mud snail, offers an excellent experimental system for examining possible effects of zero gravity on embryogenesis in general and for determining cytoplasmic factors essential for heart development in particular. Large yolk platelets accumulate gravocentrically in the vegetal pole cytoplasm of the Ilyanassa egg, displacing the nucleus to the animal pole. As a result, Ilyanassa obsoleta embryos sequester their vegetal hemisphere cytoplasm in an anuclear polar lobe at the time of their acentric first cleavage, then merge that polar lobe cytoplasm with just one of the daughter cells, and eventually form their hearts (as well as other tissues) from the cellular descendants of the polar lobe cytoplasm-containing daughter cell. If the polar lobe cytoplasm is divided equally between the two first cleavage daughter cells, morphogenesis, especially of lobe-dependent structures, is disturbed in 40% of the embryos (Render, J., J. Exp. Zool. 253:30-37, 1990). If the polar lobe is removed from the embryo at the time of first cleavage, developing embryos fail to form a beating heart, eyes, statocysts, intestine, operculum, or external shell, although they do form extensive muscle tissue (Atkinson, J.W., J. Morphol. 133:339-352, 1971; Int. J. Invertr. Reprod. Dev. 9:169-178, 1986). To examine the role of cytoplasmic factors in heart muscle development, a rapid, sensitive, and specific assay for Ilyanassa cardiac myocyte differentiation must be developed. Vertebrate heart muscle cells express heart-specific isoforms of the contractile protein myosin-II that can be distinguished from other myosin isoforms by the nucleic acid/amino acid sequence of its amino-terminal globular head region (Bement, W.M., Hasson, T., Wirth, J.A., Cheney, R.E., and Mooseker, M.S. Proc. Natl. Acad. Sci. USA 91:6549-6553, 1994). As a first step in determining whether Ilyanassa obsoleta adult hearts express a heart-specific myosin isoform that may serve as a heart-specific marker for developmental studies, reverse transcriptase-polymerase chain reaction (RT-PCR) studies using degenerate primers for myosin-II were carried out on whole adult Ilyanassa obsoleta. For comparison, a non-heart-specific Ilyanassa probe was developed using degenerate primers for a sodium-proton antiporter.

Total cell RNA was isolated from a whole adult snail using the Promega RNAgents Total RNA Isolation System. Analysis of the RNA on an agarose/ formaldehyde gel revealed intact ribosomal RNA. Degenerate probes for myosin-II were prepared against the 5' ATP binding site (5'-GARTCNGGNGCNGGNAARAC-3') and a 3' site of unknown function (5'-RTGRTTRAANARYTGYTG-3'), based on amino acid sequences conserved from C. elegans to rat, and synthesized by Genemed Biotechnologies, Inc., South San Francisco, CA 94080. Degenerate 5' and 3' oligonucleotide probes for the crab sodium-proton antiporter were generously supplied by Dr. David Towle. Ilyanassa

RNA was converted to cDNA using the GibcoBRL Superscript Preamplification System for First Strand cDNA Synthesis, and PCR was subsequently conducted using the GibcoBRL PCR Reagent System with 40 amplification cycles of 94º for 45 sec, 42º for 30 sec, and 72º for 90 sec in a Perkin Elmer Thermocycler. The myosin-II probes yielded an expected band of approximately 925 bp, whereas the sodium-proton antiporter probes yielded an expected band of 710 bp, when separated on agarose gels. These bands were then excised from the gel, cleaned using the Ambion GeniePrep Kit, ligated into gGEMt, transfected into DH5α cloned using blue-white selection, purified by 5 Prime-3 Prime Perfectprep, and sequenced by the KSU Biotechnology Sequencing Center.

Comparison of the Ilyanassa obsoleta myosin cDNA sequence with other known sequences revealed that it is most closely related to Scallop myosin-II, and that it contains sequence identities to both the striated and the smooth muscle myosin isoforms in the critical exon 5-exon 6 region that defines the difference between these two isoforms in the Scallop (Nyitray, L., Jancsó, A., Ochiai, Y., Gráf, L., and Szent-Györgyi, A.G. Proc. Natl. Acad. Sci. USA 91:12686-12690, 1994). The second most closely related myosins are the chicken and rabbit embryonic sarcomeric myosin-IIs, and the third most closely related myosins are the mouse, rat, and human alpha cardiac myosin-IIs. Current studies are underway using degenerate primers to the more 3' myosin actin-binding site to extend knowledge of the Ilyanassa myosin-II sequence to include that region. In addition, other myosin degenerate probes, shown by Blement et al., 1995, to reveal multiple additional myosin class isoforms in human and porcine cells, will be used to investigate the existence of other myosin isoforms in Ilyanassa. Comparison of the Ilyanassa obsoleta sodium-proton antiporter cDNA sequence with other known sequences revealed that it is most closely related to the rat antiporter, followed by the chinese hamster and the crab antiporters. Upon further confirmation, these Ilyanassa sequences will be submitted to Genbank, and will represent the second and third Ilyanassa sequences present in Genbank (the first is for RNA polymerase II: U10338). With this myosin-II sequence known, we are positioned to isolate Ilyanassa adult hearts and to determine, using these and other degenerate myosin probes, whether Ilyanassa hearts make tissue-specific myosin isoforms that may be used for following heart development in lobed and lobeless embryos. Research supported by NASA-NSCORT NAGW-2328.

DIRECT SEQUENCING OF PCR-AMPLIFIED Na^+/H^+ ANTIPORTER cDNA FROM THE BLUE CRAB Callinectes sapidus

Carolyn R. Newton[1], Judi A. Tilghman[2], and David W. Towle[2]

[1]Department of Biology, Kalamazoo College, Kalamazoo, MI 49006

[2]Department of Biology, Lake Forest College, Lake Forest, IL 60045

Sodium uptake by euryhaline and freshwater animals is thought to be driven by basolateral Na^+/K^+-ATPase in the absorbing epithelium (usually gills). However, at the apical membrane of these epithelial cells, it remains controversial whether amiloride-sensitive sodium channels or Na^+/H^+ antiporters are responsible for the initial uptake step. Evidence for both pathways has been obtained in closely-related species and even within a species. To distinguish between the two alternative pathways, we have begun to identify and characterize candidate sodium transporters in gills of euryhaline crabs, using molecular biological techniques.

We have determined the cDNA sequence of a putative Na^+/H^+ antiporter from gills of the green shore crab Carcinus maenas (Towle and Wu, Bull. MDIBL 33:122-123, 1994) and have described its high level of mRNA expression in the gills of this animal (Towle et al., Bull. MDIBL 34:64-65, 1995). Work currently underway is examining the expression of the antiporter mRNA with respect to salinity, following the hypothesis that acclimation to reduced salinity may lead to transcriptional activation of the antiporter gene. The green shore crab, although it is a modest osmoregulator in low salinities, cannot match the osmoregulatory ability of a related species, the blue crab Callinectes sapidus. The blue crab can penetrate environments which are essentially freshwater, and may be expected to demonstrate an enhanced ability to activate the necessary genes. We thus set out to characterize the Na^+/H^+ antiporter in gills of the blue crab using molecular techniques based on our knowledge of the Carcinus antiporter sequence.

We extracted total RNA from gills of Callinectes sapidus using a modification of the Chomczynski and Sacchi technique (Anal. Biochem. 162:156-159, 1987) (Promega Corporation). The poly-A messenger RNA in this total RNA preparation was reverse transcribed to cDNA using SuperScript II reverse transcriptase (Gibco-BRL) and oligo-dT as the primer. A segment of the putative Na^+/H^+ antiporter was then amplified from the cDNA template using degenerate oligonucleotide primers (3F and 4R, Towle et al., Bull. MDIBL 34:66-67, 1995) which were designed to produce an approximately 700-base-pair product in a successful polymerase chain reaction (PCR) amplification of the antiporter sequence. The amplification was accomplished using an MJ Research thermocycler equipped with a heated lid and programmed to first permit a "hot start" for the addition of taq polymerase (Boehringer-Mannheim), followed by 35 cycles of 92°C (60 sec), 45°C (60 sec), and 72°C (120 sec), with a final elongation step at 72°C (5 min).

Amplification products were electrophoresed on 0.8% agarose gels and visualized by ethidium bromide staining and ultraviolet transillumination. A single strong band at approximately 700 bp was removed from the gel with a sterile scalpel and its DNA was extracted using the Gene-Clean kit. Direct dideoxynucleotide sequencing of the purified PCR product was accomplished using oligonucleotide 4R as primer (PCR Products Sequencing Kit, USB), followed by acrylamide gel electrophoresis of the ^{35}S-labelled products and autoradiography.

```
                1       10      20      30      40      50      60
CALLINECTES    TTGTCATCCTCACCATCCTCTTCTGCACCATCTACAGGATACTTGGTGTGTTGATCTTCA
               ************************* ******* * ** ** ** *** * ******
CARCINUS       TTGTCATCCTCACCATCCTCTTCTGCTCCATCTACCGTATTCTAGGAGTGCTCATCTTCA
                1630    1640    1650    1660    1670    1680

                70      80      90      100     110     120
CALLINECTES    GTGTATTGTGTAACAGGTTCCGGGTGAAGAAGATCGGCTTTGTTGACAAGTTCGTGATGT
               * *    * ******** ***** ** ********** ***** ******* ******
CARCINUS       GCGCGGTATGTAACAGATTCCGCGTCAAGAAGATCGGTTTTGTGGACAAGTTTGTGATGA
                1690    1700    1710    1720    1730    1740

                130     140     150     160     170     180
CALLINECTES    CCTACGGTGGTCTGAGAGGAGCTGTTGCCTTTGCTCTTGTCATCACCATCAACCCAGAGC
               ***** ** ** ****** ** ********** ****************** *
CARCINUS       GTTACGGAGGGTTGCGAGGAGCGGTGGCCTTTGCTCTCGTCATCACCATCAACCCAATCC
                1750    1760    1770    1780    1790    1800

                190     200     210
CALLINECTES    ACATCCCTCTCCAGCCCGTGTTCCTCACTGCCACT
               ******* ********* ******* ***** ***
CARCINUS       ACATCCCACTCCAGCCCATGTTCCTTACTGCTACT
                1810    1820    1830
```

Figure 1. cDNA sequence of putative Na$^+$/H$^+$ antiporter segment produced by PCR amplification of blue crab (Callinectes sapidus) gill cDNA, using degenerate oligonucleotide primers for both amplification and direct sequencing. Alignment with corresponding segment of previously determined Na$^+$/H$^+$ antiporter from green shore crab (Carcinus maenas) (Towle and Wu, 1994) was accomplished using MULTALIN software (Corpet, 1988). Asterisks indicate agreement between the two sequences.

```
                1       10      20      30      40      50      60
CALLINECTES    VILTILFCTIYRILGVLIFSVLCNRFRVKKIGFVDKFVMSYGGLRGAVAFALVITINPEH
               ******* ***********  ********************************** *
CARCINUS       VILTILFCSIYRILGVLIFSAVCNRFRVKKIGFVDKFVMSYGGLRGAVAFALVITINPIH
                400     410     420     430     440     450

                70
CALLINECTES    IPLQPVFLTAT
               ***** *****
CARCINUS       IPLQPMFLTAT
                460
```

Figure 2. Amino acid sequence translated from the cDNA sequence of the putative Na$^+$/H$^+$ antiporter segment of blue crab (Callinectes sapidus) aligned with the previously reported, corresponding segment from green shore crab (Carcinus maenas). Asterisks indicate agreement between the two sequences.

A 215-base sequence was read from autoradiographs obtained with the Callinectes PCR product as template. When converted to its complement, the DNA sequence matched 83% of the corresponding region of putative Na^+/H^+ antiporter cDNA from the green shore crab Carcinus maenas (Towle and Wu, 1994) (Fig. 1). Translated to amino acid sequence using DNASIS software and aligned with MULTALIN (Corpet, Nuc. Acids Res. 16:10881-10890, 1988), the Callinectes sequence matched 93% of the corresponding amino acid sequence of the putative Na^+/H^+ antiporter of Carcinus (Fig. 2), representing transmembrane domains 10 and 11 of the 12 domains predicted on the basis of hydrophobicity and α-helix analysis (Jones et al., Biochemistry 33:3038-3049, 1994).

Our results indicate that the putative Na^+/H^+ antiporters of two portunid crab species are closely related at the molecular level. We have thus confirmed, by an entirely independent series of experiments, the initial characterization of the Na^+/H^+ antiporter sequence in Carcinus maenas and have extended the analysis to this second, more powerfully osmoregulatory species, Callinectes sapidus. We are now equipped to examine antiporter gene expression in both species, using oligonucleotide primers and PCR-generated probes which are specific to each species. Our work may shed light on the controversy surrounding the relative roles of sodium channels and Na^+/H^+ antiporters in the uptake of sodium ions by these animals.

Supported by a National Science Foundation grant to DWT (IBN-9407261) and a National Science Foundation Research Opportunity Award to CRN.

PARTIAL CLONING AND SEQUENCING OF THE RENAL SODIUM-D-GLUCOSE COTRANSPORTER FROM DOGFISH (<u>SQUALUS</u> <u>ACANTHIAS</u>)

[1]Alison Morrison-Shetlar, [2]Daniel Soto and [3]Brian Wolpin
[1]Biology Department, Goergia Southern University, Statesboro, GA 30460,
[2]Department of Molecular Biology and Biochemistry, Wesleyan University, Middletown CT 06459 and
[3]Pennsylvania State University, Pennsylvania, PA 16802

The proximal tubule of vertebrate kidneys re-absorbs glucose from the renal filtrate through a sodium dependent glucose cotransport system located on the apical surface of kidney epithelial cells. Low intracellular sodium concentration maintained by the Na.K-ATPase leads to the formation of a sodium gradient across the apical membrane which in turn drives the uptake of sodium into the cell. The sodium-D-glucose cotransport (SGLT) protein moves the sodium across the cell membrane, down its concentration gradient and at the same time moves glucose across the membrane against its concentration gradient. The mammalian kidney is known, through functional studies, to express a low affinity, high capacity, and a high affinity, low capacity glucose cotransporter in different locations within the renal tissue, but the difference in stoichiometry at the molecular level is not fully understood. It is the goal of this project to determine the stoichiometry for each of the transport systems, and define the glucose and sodium binding sites on the protein. We have recently cloned, sequenced and expressed the SGLT from rabbit renal tissue in <u>Xenopus</u> oocytes. Using this information we have proposed a model for the orientation of the protein in the membrane. Antibody studies have verified this model but the sodium and glucose binding sites have not been identified.

To continue this study, the spiny dogfish (<u>Squalus</u> <u>acanthias</u>) was used. This fish has been chosen because the dogfish kidney expresses only the low affinity cotransporter. We have made cDNA libraries from renal tissue of dogfish and have obtained positive clones from the library using isolated radio-labeled cDNA sequences from the known rabbit renal sequence. The 5' end of the sequence is shown in Fig 1 and is compared to that of the known rabbit kidney sequence.

A)

```
Rabbit    TGTCGTCGCCGCCGCCACGCCGCCATGGACAGCAGCACTTTGAGCCCCCT
Dogfish   gccgGcCGCtgtCttCctctCtgCATaccgAGCtaaAaTgaagGCatCtg

Rabbit    GACCACCTCCACCGCGGCCCCCCTTGAGTCCTA-TGAGCGCATCCGCAAT
Dogfish   cAtCtCCgaCAtCaaccaCgtCtccatccCaaaCTGtGgcCATCaaCAAT

Rabbit    GCGGCCGACATCTCCGTCATCGTCATCTACTTCTTGGTGGTGATGGCCGT
Dogfish   GCtGCaGAtATCagCGTcATCaTCgTGTACTTcgTttTGGTcATCGCCGT

Rabbit    CGGGCTGTGGGCTATCTTTTCCACCAATCGGGGGACGGTCGGAGGCTTCT
Dogfish   tGGaCTGTGGtCTATgTaTaggACCAAcCGtGcGACcGTcGGtGGcTaCT

Rabbit    TCTTGGCGGGTCGGAGTATGGTGTGGTGGCCGATCGGAGCCTCTCTGTTT
Dogfish   TCTTGGCgGGgaGGgacATGcgaTGGTacaCagTcGGAGCCTCaCTgTTT

Rabbit    GCCAGTAACATTGGAAGTGGCCACTTTGTGGGGCTGGCCGGGACGGGAGC
Dogfish   GCtAGTAACATcGGAAGcGGaCACTTTGTtGGttTGGCcGGcACaGGgGC
```

```
Rabbit     TGCTTCAGGCATTGCCACTGGGGGCTTTGAGTGGAACGCCCTGATCATGG
Dogfish    TGCaaacGGCcTgGCCgtcGGtGGCTTTGAGTGGAACGCCCTGtTtgTtG

Rabbit     TGGTCGTGCTGGGCTGGGTGTTTGTCCCCATTTACATCAGGGCTGGGGTG
Dogfish    TGtTacTcCTGGGtTGGCTCTTTGTCCCagTTTACcTgAcaGCTGGGGTC

Rabbit     GTGACGATGCCAGAGTATCTGCAGAAGCGGTTTGGAGGCAAGAGGATCCA
Dogfish    aTcACGATGCCCcAaTActTaatGAAGaGGTTcGGAGGaAAccGaATCag

Rabbit     GATCTACCTTTCCATTCTGTCCCTGTTGCTCTACATTTTTACCAAGATCT
Dogfish    acTCTACCTcTCCCTcaTcTCtCTcTTaCTgTACATaTTTACCAAGATCT

Rabbit     CGGCAGACATCTTTTCCGGAGCCATCTTCATCCAGCTGACCTTGGGCCTG
Dogfish    CGGtgGACATgTTcTCCGGAGCGATCTTCATCCAaCaagCtcTGGGatgG

Rabbit     GATATCTATGTGGCCATCATTATCTTATTGGTCATCACTGGGCTCTACAC
Dogfish    aAcATCTATGTtGCagTaATTgcaTTgcTGatTATtACTtGtaTCTAtAC

Rabbit     CATCACAGGGGG
Dogfish    tATCACAGaGcG
```

B)
```
Rabbit     CRRRRHAAMDSSTLSPLTTSTAAPLESYERIRNAADISVIVIYFLVVMAV
Dogfish    paalflsAyrAKmkASaSPTsTtSpSqtvaInNAADISVIVYFVLVIMAV

           GLWAMFSTNRGTVGGFFLAGRSMVWWPIGASLFASNIGSGHFVGLAGTGAA
Dogfish    GLWSMYRTNRATVGGYFLAGRDMrWYTVGASLFASNIGSGHFVGLAGTGAA

Rabbit     SGIATGGFEWNALIMVVVLGWVFVPIVIRAGVVTMPEYLQKRFGGKRIQI
Dogfish    NGLAVGGFEWNALfVVLLLGWLFVPVYLtAGVITMPQYLmKRFGGNRIRL

Rabbit     VLSILSLLLYIFTKISADIFSGAIFIQLTLGLDIYVAIIILLVITGLYTIT
Dogfish    YLSLISLLLYIFTKISVDMFSGAIFIQqALGwNIYVAVIaLLiITcIYTI
```

Fig. 1. Comparison of the A) cDNA and B) amino acid sequence obtained from dogfish kidney cDNA and the known rabbit sequence. Identity shown by capital letters in the dogfish sequence for both the DNA and amino acid sequences and similarity in the amino acid sequence shown by underlining. Non-caps means no identity nor similarity.

The results to date suggest a 56% homology at the DNA level and a 86% homology at the amino acid level.

This study will allow additional experiments to be carried out to delineate further the structure and function of this protein. Site directed mutagenesis will facilitate recognition of important functional sites on the protein. The observed functional changes in affinity will be studied in Xenopus oocytes to determine why and how affinity and stoichiometric changes occur. The immunohistochemical studies will indicate if the difference in function is due to a change in the topography of the protein in the membrane. A change in function may affect the internal environment of the cell (change in Na levels transported) and could be related to volume regulation. Additionally, regulation of this protein within the kidney cell is not fully understood and may be determined in the isolated oocyte system where large amounts of the messenger RNA injected increase expression to observable levels. Support for AIMS and BMW from Burroughs-Welcome, and from the NSF-REU No. 93-22221 grant for DS.

UPREGULATON OF THE Na-K-Cl COTRANSPORTER PROTEIN IN SALTWATER ADAPTATION OF THE RAINBOW TROUT, SALMO GAIRDNERI

Rachel Behnke, Susan Brill, Mrinilini Rao[1], and Biff Forbush
Department of Cellular and Molecular Physiology
Yale University School of Medicine, 333 Cedar St. New Haven, CT 06510
[1]University of Illinois at Chicago, 901 South Wolcott St.
Chicago Illinois, 60612-7342

Anadramous fish such as the trout undergo dramatic physiological and biochemical adaptations during the transition from a freshwater to a seawater environment. The kidneys of most marine teleosts cannot produce a concentrated urine, and extrarenal organs are known to be the major means of water absorption and salt elimination in fish adapted to seawater (D. Evans, The Physiology of the Fishes, Chapt. 11. CRC Press,1993). The intestinal epithelium is the most important site for Na, Cl, and fluid absorption in the trout (F. Conte, in Fish Physiology Vol I, Chapt. 3, Hoar and Randall eds, 1969), whereas the gill of the seawater teleost, and the chloride cell in particular, is a major site of Na+Cl secretion (K. Foskett and C. Scheffey, Science 215:164,1982). Chloride cells are known to proliferate and enlarge when fish are exposed to water of increasing salinity (K. Foskett, J. Exp. Biol. 93:209, 1981). In both absorptive and secretory processes, the Na-K-Cl cotransporter is thought to play a central role in Cl movement (M. Musch et al., in Meths. of Enzymol. 192: 746, 1990; J. Zadunaisky, in Fish Physiology, Chapt. 2, Hoar and Randall eds,1984).

Here, we test the hypothesis that an increase in the amount of the Na-K-Cl cotransporter is an important component of the saltwater adaptive response in anadramous fish. In a preliminary experiment (M. Rao and B. Forbush, unpublished) in which one of a pair of freshwater trout was acclimated to seawater, immunodetectable cotransporter greatly increased in gill, but not the kidney, of the adapted fish relative to the freshwater control. In the present investigation we gradually acclimated a set of fish to a saltwater environment and collected several tissues for protein and RNA analysis in order to observe how cotransporter expression may change in different tissues during adaptation.

Individual Salmo gairdneri, one to two years of age, were acquired from fresh water stocks at Sea Run Holdings, Maine in early September. A cohort of 27 fish were maintained, one subset (n=11) in fresh water over 7 days. The other subset (n=18) was slowly adapted to saltwater, with one day at 25%, three days at 50%, 2 days at 75% seawater, and then seven days in 100% seawater. Freshwater was obtained from the town water supply and aerated for 1-2 days before addition to aquaria. The water was continuously aerated and changed by 50% volume every day, except the 100% seawater, which was free-running. Nine tissues (heart, brain, gill, intestine, kidney, liver, stomach, skeletal muscle) were dissected. One portion of each tissue was flash frozen in liquid nitrogen for subsequent protein analysis, and the other homogenized in guanidinium isothiocyanate solution and then frozen for later RNA quantification.

We first processed tissue from gill, intestine, and kidney, focusing on the animals adapted for the longest period to see if a change in Na-K-Cl cotransporter expression was detectable between the end points of the experiment. Cell membranes were prepared from 3 fish taken at days 5 through 7 in 100% seawater and from 3 fish maintained in freshwater. Figure 1 is a Western blot of gill membrane proteins from those individuals. In all cases observed, the saltwater-adapted gill expressed between 5-and greater than 20-fold more NaKCl cotransporter protein than the freshwater controls. The intestine of the saltwater-adapted fish (data not shown) showed a similar increase in expression of the cotransporter, though the maximum amount of cotransporter was generally lower than that of the gill from the same fish. In the kidney membrane samples, there were small detectable amounts of cotransporter protein, but no significant differences in levels of this protein between adapted and control fish. These results further implicate the Na-K-Cl cotransporter as a key component in the extrarenal adaptive response of an anadramous fish to a salt water environment.

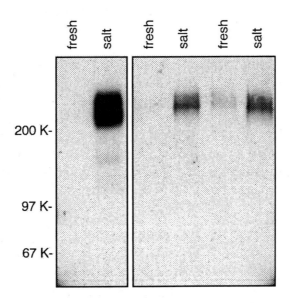

FIGURE 1. Western Blot of trout gill membranes. Samples of comparable protein were separated by PAGE and transferred to Immobilon membrane. These transfers were probed with the monoclonal antibody, T4, raised to the NaKCl cotrasport protein (Lytle et al. Am. J. Physiol. 269:C1496, 1995). The left and right panels group membrane samples prepared on different days. Protein mass is indicated on the left.

Supported by NIH grant DK-47661

OCCLUSION OF IONS BY THE Na-K-Cl COTRANSPORTER IN THE INTACT RECTAL GLAND OF THE SPINY DOGFISH, SQUALUS ACANTHIAS

Biff Forbush and Andrea Pernak
Yale University School of Medicine, 333 Cedar St., New Haven, CT 06510

It is thought that in the process of moving ions across the membrane, a transport protein may go through conformational states in which the solute molecule is hidden within bilayer, inaccessible to solutions on both sides of the membrane. The existence of such "occluded" states has been clearly demonstrated for the Na pump (Glynn, I.M., and Richards, D.E. , J. Physiol. (Lond.), 330, 17-43, 1982; Forbush, B. III , J. Biol. Chem., 262, 11104-11115, 1987) and it has been proposed that occlusion may be a general feature of the operation of transporters (Forbush, B. III in The Na$^+$,K$^+$-pump, Skou, J.C., Norby, J.G., Maunsbach, A.B., and Esmann, M. eds., 229-248, Alan R. Liss, Inc., New York, 1988).

One of the hallmarks of solute occlusion is that the behavior of the protein may be affected by the presence of the occluded molecule, and it may thereby "remember" that a solute has been present long after that solute has been removed from solution. Using membranes prepared from shark rectal gland, we have previously reported that following inhibition of the Na-K-Cl cotransporter with [^{3}H]benzmetanide and removal of the inhibitor, the rate of <u>dissociation</u> of [^{3}H]benzmetanide depends on the ions that were present during binding, not on the composition of the medium during dissociation (Forbush, B. III, and Haas, M. , Biophys. J., 55, 422a, 1989); similar observations have recently been reported in membranes from rabbit parotid gland (Moore, M.L., George, J.N., and Turner, R.J. , Biochem. J., 309, 637-642, 1995).

In the present study we examined the rate of dissociation of [^{3}H]benzmetanide from Na-K-Cl cotransporters in the intact perfused gland (cf. Forbush, B. III, Haas, M., and Lytle, C. , Am. J. Physiol., 262, C1000-C1008, 1992). In each experiment we stimulated the rectal gland with VIP, and then perfused for a 3 min period with a solution containing Rb$^+$, NH$_4^+$, or Cs$^+$ in substitution for K$^+$; during this time we introduced [^{3}H]benzmetanide for a 1 min binding period. Excess radioactivity was then removed from the gland in a 10 minute perfusion period with shark Ringer's, following which we examined the amount of [^{3}H]benzmetanide which was released into the venous effluent at 1-5 min intervals for up to 5 hours in shark Ringer's. During the entire dissociation period, the perfusion solution contained 500 μM furosemide to prevent newly released [^{3}H]benzmetanide from rebinding to uninhibited transporters. VIP was also included to insure that transporters remained in an activated state, although similar dissociation rates were observed in the absence of VIP.

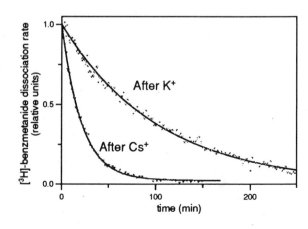

The result of a typical pair of experiments is illustrated in the figure. As shown here for Cs^+, there was a dramatic dependence of the $[^3H]$benzmetanide dissociation rate on the identity of the ion that had been present during association of $[^3H]$benzmetanide, even though the dissociation was monitored in the same solution in both experiments. Comparing K^+ substitutes (in three experiments each, except for NH_4^+), the influence on the subsequent dissociation rate of $[^3H]$benzmetanide was: Cs^+ > NH_4^+ > Rb^+ > K^+. The rate of dissociation was also 10-fold faster following binding in SCN compared to Cl. The absolute value of the rate constants (here 0.0084 min^{-1} following K^+, .042 min^{-1} following Cs^+) was about 2.5 fold lower than previously reported in isolated membranes, although the dependence upon ions was similar (Forbush, B. III, and Haas, M. , op cit). We presume this difference reflects the temperature difference between the two sets of experiments -- 15℃ here compared to 20℃ in the earlier experiments. Continuous perfusion proved to be a reproducible method for obtaining the dissociation rate: the SEM in triplicate determinations of dissociation rates was in each case less than ±4%.

These results show that the Na-K-Cl cotransporter can "remember" for several hours the ions which were present during a brief exposure to $[^3H]$benzmetanide. It seems most likely that this phenomenon occurs because the ions are trapped in the translocation pocket when the inhibitor is bound and that they can not escape until benzmetanide is released. A similar situation has been found with regard to ouabain binding by the Na pump (Forbush, B. III , Curr. Top. Membr. Trans., 19, 167-201, 1983). However unlike the Na pump, it has not yet been possible to obtain evidence that the uninhibited Na-K-Cl cotransporter occludes ions.

Supported by NIH grant DK-47661.

IMMUNOHISTOCHEMICAL LOCALISATION OF A NaPi 2-COTRANSPORT SYSTEM IN THE KIDNEY AND INTESTINE OF WINTER FLOUNDER (<u>PLEURONECTES AMERICANUS</u>)

P. Herter[1], M. Elger[2], B. Kohl[1], L. Renfro[3], H. Hentschel[1], R.K.H. Kinne[1], and A. Werner[1]

[1]Max-Planck-Institut für molekulare Physiologie, D-44026 Dortmund, FRG
[2]Institut für Anatomie und Zellbiologie I, Universität, D-69120 Heidelberg, FRG
[3]University of Connecticut, Dept. of Physiology and Neurobiology, Storrs, USA

After the cloning of a renal Na/Pi 2-cotransport system from winter flounder (Werner A, et al. Am. J. Physiol. 267: F311-317, 1994) we recently investigated by immunohistochemistry the distribution of this transport protein along the nephron. A further question of interest was whether the renal Na/Pi-transporter is also present in flounder intestine.

Winter flounder were caught in the Gulf of Maine in November 1994 and July 1995. Kidney and intestine were fixed by dripping a chilled mixture of 2% paraformaldehyde and 0.5% picric acid in 80% ethanol on the exposed organs or by perfusion with 4% paraformaldehyde. Blocks of tissue were dehydrated via graded ethanol and embedded in paraffin. For immunodetection deparaffinized sections (7 µm) were incubated with two antisera raised in rabbits against different partial sequences of the transport protein. Blocking of unspecific binding sites was performed for 15 min. with 50 mM glycine in PBS and 20 min. in PBS containing 5% goat serum, 0.2% gelatine and 0.5% BSA. Incubations with preimmune serum served as controls. Primary antisera and preimmune serum were diluted 1:400 in PBS with 0.2% gelatine and 0.5% BSA (PBG) and incubated for 2 hours with the sections. As detection system a secondary goat anti rabbit IgG antibody conjugated to Cy 3 (DIANOVA, Hamburg, Germany) diluted 1:100 in PBG was used. In double labeling experiments with <u>Lens culinaris</u> agglutinin (LCA) the lectin was diluted 1:50 in PBG.

We observed distinct binding of antisera specific for the Na/Pi -cotransport protein at the region of basolateral membranes of epithelial cells of the proximal tubule segment PII (Fig. 1a). Labelling was absent in controls with omission of the primary antisera or after incubation with preimmune serum. By electron microscopy of thin sections of flounder kidney, it was revealed that the basolateral cell membranes are greatly amplified by stacks of intracellular infoldings in the subnuclear cytoplasm (personal unpublished results). Our light-microscopic observations suggest sorting of the antigene to this region, as shown by the comma-shaped reaction product. Moreover, the NaPi transporter apparently is sorted differently in the flounder as compared to the mammalian proximal tubule (Murer H., NIPS 10:287, 1995). The identification of the two segments PI and PII on histological sections was possible after we had screened a variety of plant lectins for their affinity to flounder kidney structures (for methodology of lectin histochemistry see Hentschel et al., Bull MDIBL 34:32-35, 1995). Our results with the lectins revealed that Lens culinaris agglutinin (LCA) selectively binds to the brush border of PI in Winter flounder (Fig. 1b).

In flounder intestine labeling with the Na/Pi antisera could be detected in the brush border region and the subapical cytoplasm of the enterocytes (Fig. 2). The same sodium transport system seems to be present both in flounder kidney and intestine, however at different cell poles. The presumed presence of the transporter in basolateral membranes of the PII-segment - a predominantly secretory epithelium - suggests an involvement of the renal system in phosphate secretion whereas in intestine it might be involved in phophate absorption.

This study was funded by Max-Planck-Society and Deutsche Forschungsgemeinschaft (Travel grant to M.E.) H.H. was a recipient of an MDIBL New Investigators Award.

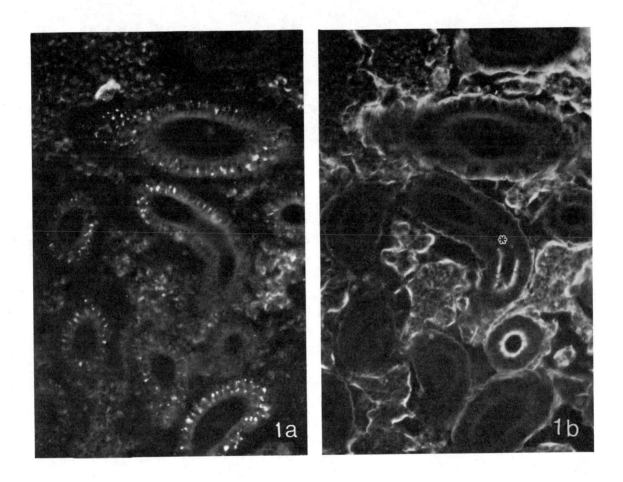

Fig. 1a Localization of the Na/Pi 2-cotransporter in the PII segment of the proximal tubule by indirect immunofluorescence. Epithelial cells of the PII-segment show immunoreaction at their basolateral membranes.

Fig. 1b. Identification of proximal tubule PI-segments by staining with FITC-LCA. Lectin binding sites are located at the brush border of PI-cells and at the endothelial cells of the venous sinuoid capillaries. A transition of PI to PII is marked with an asterisk.

Fig. 2. Localization of the Na/Pi 2-cotransporter in the intestine of the winter flounder by indirect immunofluorescence. A marked labelling is present at the brush border region and the subapical cytoplasm of the enterocytes. The goblet cells are negative.

HOMOLOGS OF AN EXTRACELLULAR CALCIUM/POLYVALENT CATION-SENSING RECEPTOR (CaR) ARE LOCALIZED TO THE APICAL SURFACES OF SPECIFIC EPITHELIAL CELLS IN ORGANS CRITICAL FOR IONIC HOMEOSTASIS IN THE ELASMOBRANCHS, <u>SQUALUS ACANTHIAS</u> AND <u>RAJA ERINACEA</u>, AS WELL AS TELEOSTS (<u>PLEURONECTES AMERICANUS</u>), (<u>ONCORHYNCHUS MYKISS</u>) AND <u>FUNDULUS HETEROCLITUS</u>

Michelle Baum[1], Hartmut Hentschel[2], Marlies Elger[3], Edward Brown[4], Steve Hebert[5] and H. William Harris[1]

[1]Div. of Neph., Children's Hospital, Boston, MA 02115
[2]Max-Planck-Instit. of Mol. Physiology, Dortmund, Germany
[3]U of Heidelberg, Instit. of Anatomy/Cell Biol., Heidelberg, Germany
[4]Endo.-Hyperten. Div. and [5]Renal Div. Brigham and Women's Hospital, Boston, MA 02115.

Molecular cloning and characterization of a cell surface receptor called the calcium/polyvalent cation sensing receptor or CaR has demonstrated that it responds to or "senses" extracellular Ca^{2+} and Mg^{2+} concentrations of 1-5 mM Ca^{2+} and 5-20 mM Mg^{2+} respectively. CaR is expressed by mammalian parathyroid and C cells as well as several tubule epithelial cells of the kidney including thick limb of Henle (TAL). CaR allows these cells to respond to alterations in serum Ca^{2+} and Mg^{2+} by modulation of intracellular signal transduction pathways (Brown, E.M. et al. New Eng. J. Med. 333:234, 1995). Recent data (Baum, M. et al. J. Am. Soc. Neph. 6: 319A, 1995) have shown that a CaR is also present on the apical membranes of rat inner medullary collecting duct where it modulates vasopressin-elicited water permeability. Since fish encounter alterations in the ambient and urinary concentrations of both Ca^{2+} and Mg^{2+} during fresh to seawater transitions, we studied both the distribution and expression of CaR homologs in marine and euryhaline species. The distribution of CaR protein was determined using an anti-CaR antiserum and immunohistochemistry of tissue sections while CaR expression was surveyed by RNA blotting using CaR specific cDNA probes.

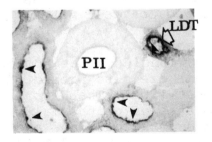

FIGURE 1: Localization of CaR in <u>S. acanthus</u> kidney tubules using anti-CaR specific antiserum. CaR protein (shown as the dark reaction product indicated by arrowheads) is present on the apical membrane of epithelial cells of the collecting duct (CD) and late distal tubule (LDT) but not PII segment (PII) responsible for Mg^{2+} secretion. (Mag 400X).

In both skate and dogfish, CaR protein was localized to the apical membranes of selected epithelial cells in kidney tubules, rectal gland and gill. In the kidney, CaR protein was confined to the CD and LDT (Figure 1). These cells are in close proximity to and receive the luminal fluid of the PII segment that is responsible for renal Mg^{2+} secretion (Hentschel, H. and K. Zierold Eur. J. Cell Biol. 63:32, 1994). The presence of CaR in these elasmobranch nephron segments suggests that CaR may play a role in

31

regulation of renal divalent metal ion secretion by sensing alterations in urinary Mg^{2+} and Ca^{2+} concentrations.

In the flounder (<u>Pleuronectes americanus</u>), CaR protein was localized to the apical membranes of epithelial cells in kidney tubules, gill, urinary bladder and intestine as well as specific regions of brain. In the fresh water trout (<u>Onchorhynchus</u>), CaR staining was present in the urinary bladder.

CaR protein was also localized to the apical membranes of epithelial cells in kidney tubules in the killifish, <u>Fundulus heteroclitus</u>. To determine if CaR expression was modulated by adaptation of <u>Fundulus</u> to either fresh or salt water, killifish collected in the local estuary were first fresh or salt water adapted for an interval of 18 days (chronic adaptation). Selected individuals from each group were then adapted to the corresponding salinity (fresh to salt; salt to fresh) for an interval of 7 days (acute adaptation). As shown in Figure 2, chronic adaptation to seawater results in an increase in steady state levels of CaR mRNA in 3 tissues. In a similar manner, we also observed increases in both the staining intensity of CaR as well as the number of epithelial cells possessing CaR in kidneys of fish adaptated chronically as well as acutely to salt water as compared to fresh water conditions.

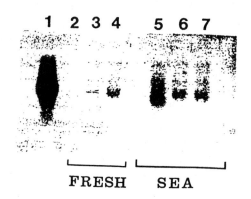

Figure 2: CaR expression in tissues of <u>Fundulus heteroclitus</u> after an 18 day interval of fresh (lanes 2-4) or sea water (lanes 5-7) adaptation. A blot containing RNA (40 μg/lane) prepared from control Xenopus kidney (lane 1) or <u>Fundulus</u> heart (containing ultimobranchial tissue) (lanes 2, 5), kidney (lanes 3, 6) and gill (lanes 4, 7) was probed with a ^{32}P-labeled Xenopus CaR cDNA, washed (0.1 X SSC, 65°C) and autoradiographed. As compared to control mRNA (lane 1), steady state levels of CaR mRNA are larger in tissues from sea water adapted fish (lanes 5-7) vs those in fresh water (lanes 2-4).

We conclude that homologs of the CaR protein present in mammals are also present in selected tissues of elasmobranch and teleost fish. The distribution of CaR protein on the apical membranes of epithelial cells in the gill, intestine, urinary bladder, rectal gland and kidney tubules as well as brain suggests the involvement of CaR in modulation of epithelial ion and water transport and perhaps endocrine function. Alterations in the steady state levels of CaR mRNA in ultimobranchial tissue, kidney and gill that accompany adaptation of <u>Fundulus</u> to either fresh or salt water further suggests a role for CaR in fish osmoregulation.

These studies were funded jointly by the Maine Aquaculture Innovation Center (MAIC) and NPS Pharmaceuticals Inc. Salt Lake City, UT.

HYDROGEN PEROXIDE REGULATES THE RYANODINE RECEPTOR AND THE Na$^+$-Ca^{2+} EXCHANGER IN RAT (RATTUS NORVEGICUS) CARDIAC MYOCYTES

Yuichiro J. Suzuki, Jose Monterubio, Barbara A. Hughes, Lars Cleemann, Darrell R. Abernethy and Martin Morad

Institute for Cardiovascular Sciences and Department of Pharmacology, Georgetown University Medical Center, Washington, DC 20007 USA

Ca^{2+} plays an important role in cardiac muscle excitation-contraction coupling. The $[Ca^{2+}]_i$ is controlled by sarcolemmal Ca^{2+} channels and transporters. Voltage-gated, dihydropyridine-sensitive L-type Ca^{2+} channels serve as the first step in signaling of Ca^{2+} release from the sarcoplasmic reticulum (SR) (Nabauer et al., Science 244: 800-803, 1989). The Ca^{2+} extrusion process is controlled primarily by Na$^+$-Ca^{2+} exchanger which transports Ca^{2+} out of the cell against its concentration gradient utilizing the favorable gradient of Na$^+$ ions.

Reactive oxygen species (ROS) such as superoxide (O$_2\bullet^-$), hydrogen peroxide (H$_2$O$_2$) and hydroxyl radicals (HO•) have been considered to be extremely reactive and inhibitory to biological molecules. Recent findings, however, showed that ROS also stimulate signal transduction pathways. For example, O$_2\bullet^-$ was found to enhance the IP$_3$-induced Ca^{2+} release from the vascular smooth muscle SR (Suzuki & Ford, Am. J. Physiol. 262: H114-H116, 1992). Therefore, ROS may be physiologically important signal transducing molecules in various tissues in diverse species including mammals, bacteria, plants and marine organisms.

ROS are implicated in myocardial ischemia-reperfusion injury, and a number of studies have examined the inhibitory effects of ROS on signal transduction components of cardiac muscle using high (mM) concentrations of ROS. In the present study, we have examined possible effects of physiological concentrations of ROS on Na$^+$-Ca^{2+} exchanger and Ca^{2+} release in isolated cardiac myocytes. We find that μM levels of H$_2$O$_2$ augment the activity of Na$^+$-Ca^{2+} exchanger and the Ca^{2+} channel-gated release of Ca^{2+} from the SR.

Ventricular myocytes were isolated from male Wistar rats using the collagenase/protease method as described (Mitra & Morad, Am. J. Physiol. 249: H1056-H1060, 1981). Whole cell clamped myocytes were dialyzed with 0.2 mM Fura-2, and voltage-dependent Ca^{2+} current (I_{Ca}) and SR Ca^{2+} release transients were simultaneously monitored (Cleemann & Morad, J. Physiol. 432: 283-312, 1991). Excitation wavelengths of 335 nm and 410 nm were used to monitor the fluorescence signals of Ca^{2+}-bound and Ca^{2+}-free Fura-2, and $[Ca^{2+}]_i$ was calculated as described by Cleemann & Morad (1991). Changes in $[Ca^{2+}]_i$ largely reflect changes in cytosolic Ca^{2+} in response to SR Ca^{2+} release (Cleemann & Morad, 1991). Since rapid application of caffeine causes intracellular Ca^{2+} release, and activates inward Na$^+$-Ca^{2+} exchange current, $I_{Na/Ca}$ (Callewaert et al., Am. J, Physiol. 257: C147-C152, 1989), SR Ca^{2+} release and $I_{Na/Ca}$ were simultaneously monitored. Extracellular solution contained (in mM): 137 NaCl, 5.4 KCl, 2 CaCl$_2$, 10 HEPES, 1 MgCl$_2$ and 10 glucose (pH 7.4). Intracellular solution contained (in mM): 110 CsCl, 30 TEA-Cl, 5 MgATP, 10 HEPES, 0.1 cAMP and 0.2 Fura-2. Cells were exposed to caffeine, H$_2$O$_2$ and/or dithiothreitol (DTT) using the rapid (<50 ms) perfusion system. The observed phenomena are presented by showing representative experimental results. Each phenomenon was reproduced in at least five different rat or guinea pig ventricular myocytes. Efficacy of Ca^{2+}-induced Ca^{2+} release was calculated by the equation, $(dCa/dt)/I_{Ca}$.

Rapid exposure of cells to μM levels of H_2O_2 resulted in suppression of I_{Ca}, but enhancement of Ca^{2+} release, suggesting increased efficiency of Ca^{2+}-induced Ca^{2+} release mechanism. Fig. 1 shows voltage-dependent (A) I_{Ca} and (B) SR Ca^{2+}-transient as membrane potential was depolarized from -80 to 0 mV. Exposure of the cell to H_2O_2 (100 μM) inhibited I_{Ca} (Fig. 1A), but enhanced intracellular Ca^{2+} transients (Fig. 1B). According to the Ca^{2+}-induced Ca^{2+} release theory, a decrease in I_{Ca} would have resulted in decreased Ca^{2+} release. Thus, the enhancement of Ca^{2+} transients by H_2O_2 suggests increased efficacy of Ca^{2+} release mechanism gated by I_{Ca}. The data suggest that H_2O_2 increases the efficacy of Ca^{2+}-induced Ca^{2+} release by 2-fold.

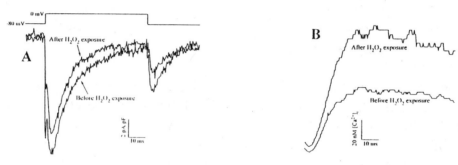

Fig. 1 Voltage-dependent (A) I_{Ca} and (B) Ca^{2+}-release before and after the H_2O_2 exposure

Consistent with this idea, in vitro studies of ryanodine receptor incorporated in lipid bilayers have shown that H_2O_2 enhances mean open time of the ryanodine receptor (Boraso & Williams, Am. J. Physiol. 267: H1010-1016, 1994). Figure 2 further supports this observation as acute exposure of the cell to H_2O_2 (100 μM) causes a rise in $[Ca^{2+}]_i$ and also activates $I_{Na/Ca}$. These Ca^{2+} transients appear quite similar to those induced by 5 mM caffeine.

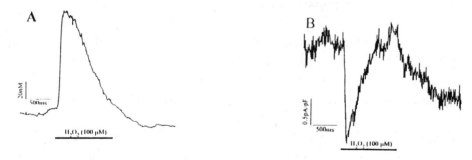

Fig. 2 H_2O_2-induced (A) Ca^{2+} release and (B) Na^+-Ca^{2+} exchange current

We conclude that, in intact rat cardiac myocytes, H_2O_2: (1) enhances the efficacy of Ca^{2+}-induced Ca^{2+} release; and (2) induces activity of the Na^+-Ca^{2+} exchanger. Our observations suggest that H_2O_2 may regulate Ca^{2+}-signaling in cardiac myocytes.

This work was supported by NIH Grant HL16152 and a grant-in-aid from AHA/Maine Affiliate to M.M. and Georgetown Univ. Medical Center Research Starter Grant Award to Y.J.S.

THE INHIBITORY EFFECT OF ATP ON CHLORIDE SECRETION BY THE RECTAL GLAND OF SQUALUS ACANTHIAS IS NOT MEDIATED BY THE ADENOSINE A1 RECEPTOR.

Patricio Silva,[1] Richard Solomon[1], Katherine Spokes[2], Kathrina Mooney[3], Dalaina Gumbs[4], Robin Fraley[5], and Franklin H. Epstein[2]
[1]Department of Medicine, Harvard Medical School and New England Deaconess Hospital and Joslin Diabetes Center, Boston, MA 02215.
[2]Department of Medicine, Harvard Medical School and Beth Israel Hospital, Boston, MA 02215.
[3]Oglethorpe University, Atlanta, GA 30319.
[4]High School for Environmental Studies, New York, NY 10019.
[5]Mount Desert Island High School, Mount Desert, ME 04660.

We have previously shown that ATP and related nucleotides inhibit the secretion of chloride by the isolated perfused rectal gland of the shark (Silva, P., et al., Bull. Mt. Desert Isl. Biol. Lab. 34:49, 1995 and Silva, P., et al., Bull. Mt. Desert Isl. Biol. Lab. 33:75, 1994). The effect of ATP is not mediated by adenosine because it can be elicited by ATP analogs that are not hydrolyzable to adenosine and it is also evoked by the pyrimidine based nucleotide UTP that does not yield adenosine upon hydrolysis. These nucleotides may exert their inhibitory effect through inhibitory adenosine receptors present in the rectal gland. The following experiments were done to ascertain this possibility.

Shark rectal glands were perfused as described in Silva P, et al. Methods Enzymol. Vol 192:754-66, 1990. The glands were stimulated to secrete chloride with forskolin 10^{-6}M. The adenosine analog 8-cyclopentyl methyl xanthine 10^{-5}M (CPT) was used to block the inhibitory A1 adenosine receptor. Two isolated rectal glands were perfused with CPT and adenosine 10^{-6}M to test for the effectiveness of CPT to block the inhibition of chloride secretion normally seen with this concentration of adenosine. CPT blocked the effect of adenosine.

CPT did not block the inhibitory effect of ATP. Figure 1 summarizes the results. In glands stimulated to secrete chloride with forskolin 10^{-6}M, β,γ methylene ATP inhibited the secretion of chloride. In the presence of CPT, β,γ methylene ATP also inhibited the secretion of chloride. Thus, CPT, that blocks the inhibitory effect of adenosine, does not prevent the inhibitory effect of β,γ methylene ATP. We conclude from these results that the effect of ATP to inhibit the secretion of chloride by the rectal gland is not mediated by the inhibitory adenosine receptor.

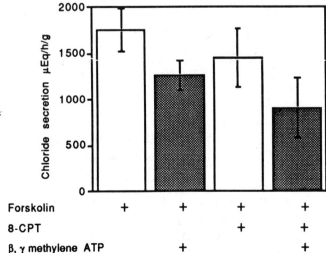

Figure 1. The effect of CPT on the inhibition of the secretion of chloride by isolated perfused rectal glands by β,γ methylene ATP. CPT does not prevent the inhibitory effect of β,γ methylene ATP. Values are mean±SEM, n=5 for glands perfused without 8-CPT and 6 for those perfused with 8-CPT.

Supported by grants from The American Heart Association: Maine Affiliate, NIEHS ESO3828, and NIH AM18098 and NSF ESI-9452682 for DG and RF.

THE SECRETION OF CHLORIDE BY THE RECTAL GLAND OF THE LITTLE SKATE, RAJA ERINACEA.

Patricio Silva,[1] Richard Solomon[1], Katherine Spokes[2], and Franklin H. Epstein[2]
[1]Department of Medicine, Harvard Medical School and New England Deaconess Hospital and Joslin Diabetes Center, Boston, MA 02215
[2]Department of Medicine, Harvard Medical School and Beth Israel Hospital, Boston, MA 02215

The rectal gland of Raja erinacea secretes chloride in response to dibutyryl cyclic AMP plus theophylline, or forskolin (Fletcher, L. et al., Bull MDIBL 23:12, 1983). In those experiments, vasoactive intestinal peptide (VIP) induced a 50% increase in the secretion of chloride but this increase was not statistically significant. The present experiments were performed to determine whether C-type natriuretic peptide (CNP) stimulated the secretion of chloride by this gland.

Glands were perfused as previously described (Fletcher, L. et al., Bull MDIBL 23:12, 1983) with shark Ringer's using a Harvard infusion pump at 1.1 ml/min. CNP (human, porcine) was dissolved in shark Ringer's. Boiled extracts of skate heart and intestine were prepared in phosphate buffer, lyophilized, suspended in distilled water and dissolved in shark Ringer's for use. Samples of heart and intestinal extracts were assayed for adenosine, and no adenosine was found.

The results are shown in Table I. Forskolin stimulated the secretion of chloride 10 fold, an increase of the same magnitude as that previously observed. CNP stimulated the secretion of chloride 40 fold. The effect of CNP was rapid and sustained. Both heart and intestinal extract stimulated the secretion of chloride, but their stimulation was of much smaller magnitude. In addition, in two glands, Scyliorhinin II, a peptide that stimulates the secretion of chloride by the rectal gland of Scyliorhinus canicula, was found to have no stimulatory effect. A similar lack of effect of Scyliorhinin II was found in the rectal gland of S. acanthias.

TABLE I

	Basal	Exp1	Exp2	n	p
Forskolin 10^{-6}M	177 ± 95	617 ± 134	1832 ± 371	12	0.01
CNP 10^{-8}M	29 ± 29	1194 ± 553	1212 ± 389	4	0.05
Heart extract	49 ± 49	324 ± 141		6	0.05
Intestinal Extract	186 ± 39	431 ± 84	540 ± 236	3	0.05

Basal, Exp1, and Exp2 are consecutive collection periods. Values are µEq of chloride secreted per hour per gram weight ± SEM. Statistical analysis was done using standard "t" test between Basal and Exp1 periods.

These results confirm the previous finding that the rectal gland of the little skate is stimulated by forskolin. The stimulatory effect of CNP contrasts with the previous observation that VIP 3×10^{-6} M has only a modest stimulatory effect if one at all. The difference in the response to VIP and CNP is striking when compared with that of the rectal gland of S. acanthias where CNP and VIP have similar effects at equimolar concentrations. The present finding suggests that in the skate, CNP may have a direct effect independent of VIP, an observation that needs further investigation. The stimulation observed with heart extract is probably due to the presence of CNP or a related peptide in the skate heart. The peptide(s) present in the skate intestine responsible for the stimulatory effect are not known.

Supported by grants from The American Heart Association: Maine Affiliate, NIEHS ESO3828, and NIH AM18098

AMMONIUM CHLORIDE INHIBITS CHLORIDE SECRETION IN THE RECTAL GLAND OF SQUALUS ACANTHIAS

Richard Solomon[1], Patricio Silva[1], Robin Fraley[2], Katrina Mooney[3] Amanda Hoche[4], Dalaina Gumbs[5], Sam Solomon[6], James Boyer[7], Jeffrey Matthews[8], and Franklin Epstein[9]

[1]Department of Medicine, New England Deaconess Hospital and Joslin Diabetes Center, Harvard Medical School, Boston, MA, [2]Mt. Desert Island High School, Bar Harbor, ME, [3]Ogilvie College, Atlanta, GA, [4]Stanford University, Palo Alto, CA, [5]High School for Environmental Studies, New York, NY, [6]Park School, Brookline, MA, [7]Department of Medicine, Yale School of Medicine, New Haven, CT, [8]Department of Surgery, Beth Israel Hospital, Harvard Medical School, Boston, MA [9]Department of Medicine, Beth Israel Hospital, Harvard Medical School, Boston, MA

Recent studies in intestinal and renal epithelia have indicated that NH_4 can substitute for K in a variety of membrane transport processes including Na, K, 2Cl cotransport and Na+K ATP hydrolysis. In T84 cells, derived from chloride secreting intestinal crypt cells, NH_4 inhibits short circuit current stimulated by cAMP agonists. The following studies were performed to determine the effect of NH_4 on chloride secretion in the rectal gland, a tissue analagous to the intestinal crypt cell.

Rectal glands were perfused with shark Ringer's solution as previously described (Solomon et al., Am. J. Physiol.262: R707, 1992). Nine collection periods of 10 minutes each were obtained during constant stimulation of chloride secretion by theophylline, 2.5×10^{-4} M. After three baseline collections, the perfusate was switched to one that had ammonium chloride, 1, 5, or 10 mM, substituted equimolar for sodium chloride. After three collections during exposure to the ammonium chloride Ringer's, the perfusate was again switched back to the original shark's Ringer's without ammonium chloride. As seen in Figure 1, addition of ammonium chloride (5 mM) to the perfusate produced a statistically significant and reversible inhibition of chloride secretion.

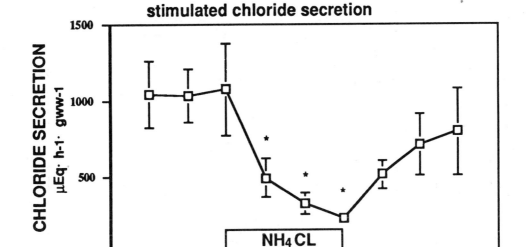

The effect of ammonium chloride on theophylline stimulated chloride secretion

Figure 1.Ammonium chloride, 5 mM, inhibited theophylline stimulated chloride secretion within 10 minutes of exposure. The effect was reversible with removal of ammonium chloride from the perfusate. * p < .05 by paired t test compared to the value at 30 minutes.

Rectal glands stimulated by theophylline were exposed to either 1, 5, or 10 mM NH_4Cl (Figure 2). Total chloride secretion was inhibited by 59% at 1mM NH_4, while at 5 and 10 mM NH_4, 73% and 71% of chloride secretion respectively was inhibited. At all concentrations, recovery of chloride secretion occurred rapidly following cessation of exposure to NH_4Cl. The inhibitory effect of 1 mM NH_4Cl was not statistically different from that of 5 mM or 10 mM NH_4Cl.

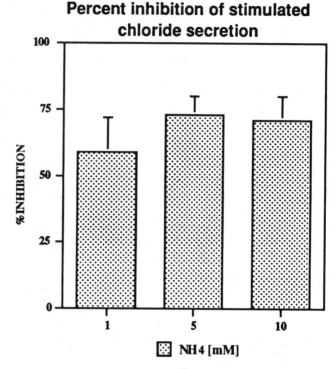

Percent inhibition of stimulated chloride secretion

Figure 2. Ammonium chloride inhibits theophylline stimulated chloride secretion.

NH_4Cl (5 mM) also inhibited chloride secretion stimulated by forskolin and genistein (Figure 3). The degree of inhibition was similar to that observed under stimulation with theophylline. Recovery following removal of the NH_4Cl was again rapid and complete.

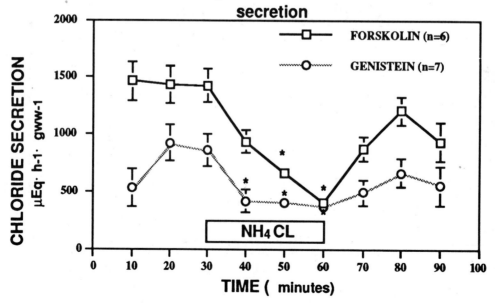

The effect of ammonium chloride on forskolin and genistein stimulated chloride secretion

38

Figure 3. Ammonium chloride, 5 mM, inhibited both forskolin and genistein stimulated chloride secretion within 10 minutes of exposure. The effect was reversible with removal of ammonium chloride from the perfusate. * p < .05 by paired t test compared to the value at 30 minutes.

NH4Cl also inhibited short circuit current in primary cultures of rectal gland cells grown to confluence on collagen supports and mounted in Ussing chambers. Ion transport was stimulated with C-type natriuretic peptide (10^{-7}M) and NH4Cl (5 mM) applied to the apical surface. Short circuit current was inhibited $50 \pm 11\%$ (n=3; data not shown).

NH4Cl is a weak acid and can affect intracellular pH either as a result of diffusion of NH3 or transport of NH4 ions into the cell. We determined the effect of NH4Cl incubation on intracellular pH of freshly prepared rectal gland tubules loaded with the pH sensitive fluorescent dye, 2'-7'-bis(carboxyethyl)-5(6)-carboxyfluorescein, BCECF. Fluorescence was monitored in a dual beam spectrophotometer and the ratio of fluorescence at 439 and 505 nm calculated. The exposure to either 5 or 10 mM NH4Cl led to transient alkalinization but pHi rapidly returned to basal levels and remained stable for at least 10 minutes (Figure 4). Nigericin that allows equilibrium between the intracellular and extracellular pH was used as a positive control for alkalinization. The transient alkalinization following exposure to NH4Cl presumably results from the diffusion of NH3 into the cells, acquisition of a H^+ to form NH_4^+ resulting in an increase intracellular pH.

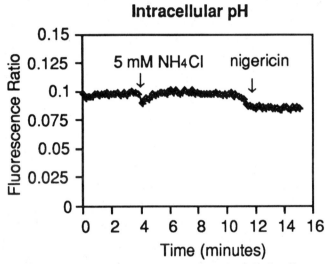

Figure 4. Intracellular pH measured with the fluorescent dye BCECF was not affected by exposure of the cells to ammonium chloride.

These preliminary observations provide evidence that NH4Cl inhibits chloride transport. The experimental design indicates that it is the NH4 ion which produces these effects. The results confirm the observations of Prasad et al. in T84 cells (J. Clin. Invest. 96:2142,1995). The inhibition does not appear to be a consequence of an effect of NH4 on intracellular pH as this remained unchanged during the period of inhibition of chloride secretion. Inhibition of both cAMP and cGMP mediated stimulation of chloride secretion suggests that the effect of NH_4^+ occurs at a distal site in the intracellular signalling cascade leading to enhanced chloride secretion, perhaps at the chloride channel itself.

The physiologic importance of these observations is unclear. In elasmobranchs, a reduced capacity (relative to mammals) to synthesize urea from NH3 results in a 10 fold higher concentration of blood NH4 than that found in mammals (Goldstein, L. personal communication). Arterial blood levels are in the order of 0.2-0.4 mM. Future studies will address whether there is an effect of such levels on chloride secretion by the rectal gland and the mechanism of the inhibitory effects reported herein.

Supported by NSF REU 9322221, NSF ESI 9452682, NIEHS ESO-3828-10, and the American Heart Association, Maine Affiliate.

RENAL MORPHOLOGY OF THE KILLIFISH, <u>FUNDULUS HETEROCLITUS</u>

Hartmut Hentschel[1], Marlies Elger[2]
[1] Max-Planck-Institut für molekulare Physiologie, D-44026 Dortmund, FRG
[2] Institut für Anatomie und Zellbiologie I, Universität, D-69120 Heidelberg, FRG

The killifish, <u>Fundulus heteroclitus,</u> is exposed to profound changes in ambient salinity in its natural habitat (tidal zone of estuaries) at regular intervals. The kidney of these fish is involved in different types of osmoregulation, as they perform (a) glomerular filtration and fluid excretion in fresh water, and (b) aglomerular fluid and electrolyte secretion in salt water. Like kidneys of other marine teleosts, the nephronic tubule consists of proximal segments only, which are connected to the archinephric duct by a collecting-tubule-collecting duct system (Hentschel and Elger, Adv. Anat. Embryol. Cell Biol. 108: 1 - 151, 1987). By virtue of the scarcity of interstitial tissue and the ease of dissection of single tubules the proximal tubules of killifish provide an excellent system for the study of various active transport mechanisms in living renal tubules (Miller et al., Am. J. Physiol. 246: R882 - R890, 1993; Miller, Am. J. Physiol. 269: R370 - R379, 1995).

We studied the arrangement and ultrastructure of proximal tubules with light- and electronmicroscopy as a basis for the investigation of Mg-transport in epithelial cells and for identification of renal cells in immuno-electronmicroscopy. Whole fish were perfused via the heart with teleost Ringer's containing 2% dextran (2 to 5 min), this perfusion was followed by perfusion fixation with a glutaraldehyde-paraformaldehyde mixture. Semithin (0.5 μm) and thin (50-70 nm) Epon sections were prepared. Renal tissue was located in two thin straps lying retroperitoneally along the vertebrate column. Cranially of the kidney straps, in the vicinity of the heart, the head kidney made up a pair of compact masses, which contained only a few tubules embedded in hematopoetic tissue. At the caudal end of the body cavity, the archinephric ducts united to form the urinary bladder. A cross section through a lobule of one kidney strap is shown in fig. 1. In the electron microscope, the proximal tubule cells displayed an apical brush border of microvilli, straight lateral cell borders, and an extremely elaborate system of basolateral infoldings, as is characteristically found in teleost proximal tubule cells (Hentschel and Elger, in Kinne et al. (eds.) Structure and Function of the Kidney, Karger, Basel 1989, pp. 1.72). In <u>Fundulus</u> these infoldings are present in parallel stacks of more than ten lamellae, resulting in large areas of sectioned membranes between the numerous and large mitochondria on electronmicrographs.

The two portions of proximal tubule differed by the organization of the apical cytoplasmic zones. The proximal tubule PI was characterised by an elaborate endocytic apparatus consisting of coated pits, coated vesicles, early endosomes, lysosomal apparatus and dense apical tubules (fig. 2). The microvilli of PII were generally longer than those of the subsequent segment PII. In the proximal tubule PII the endocytic apparatus was lacking. Instead, clear vesicles (diameter 200 to 400 nm), resembling vesicles in the trans-Golgi region, were seen in the apical cytoplasm. Frequently they were seen merging with the apical cell membrane at the base of the microvilli (fig. 3). These vesicles had no coat and therefore differed from those in the proximal tubule segment PI. In addition, small electron dense particles, which indicated filtered dextran (Christensen and Maunsbach, Virchows Arch. (B) 37: 49 - 59, 1981) were only found in apical pits and vesicles of PI. We believe that the majority of the apical vesicles in P II are exocytic. Exocytic vesicles of a similar size and morphology were described in dogfish PII. In this marine cartilaginous fish the vesicles of PII contained high concentrations of magnesium (Hentschel and Zierold, Europ. J. Cell Biol. 63: 32 - 42, 1994).

In conclusion, the proximal tubule PII cells were similar to mammalian proximal tubule cells by the ultrastructural organization of the apical cytoplasmic zone. In contrast, PII cells displayed morphological specializations which are thought to be correlated to the function of fluid and salt secretion (NaCl) in aglomerular urine formation, as well as divalent ion secretion.

The study was supported by Max Planck Society, the Deutsche Forschungsgemeinschaft (EL 92/1-1). H.H. was a recipient of an MDIBL New Investigators Award.

Figure 1. Cross section (0.5μm thick) through a lobule of the kidney of killifish, Fundulus heteroclitus. The dorsal aorta (DA), a renal artery (RA), a glomerulus, a collecting tubule (CT), and an archinephric duct (AD) are seen in cross section. The convoluted proximal tubule is present in different section profiles (PI, proximal tubule segment I, PII, proximal tubule segment II). Methylene blue-Azur A staining. x 250.

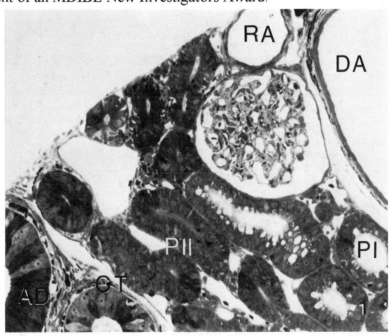

Figure 2. Electronmicrograph of the the apical zone of a PI cell. Perfusion with glutaraldehyde-paraformaldehyde after perfusion with teleost Ringer's containing 2% dextran 6000. The apical cytoplasm is endowed with a complement of endocytic organelles (coated pits, early endosomes) and dense tubules. Coated pits (arrow) and early endosomes contain dense particles, presumably clusters of dextran molecules. LY lysosome-like inclusions.
x 23 000

Figure 3. Electronmicrograph of the the apical zone of a PII cell. The apical cytoplasmic zone is devoid of an endocytic apparatus, as demonstrated by the lack of dextran particles in the clear vesicles. The cytoplasm is filled with tubular profiles, probably belonging to the smooth ER. Smooth clear vesicles are present. Arrows point to vesicles merging with the apical cell membrane. x 35 000

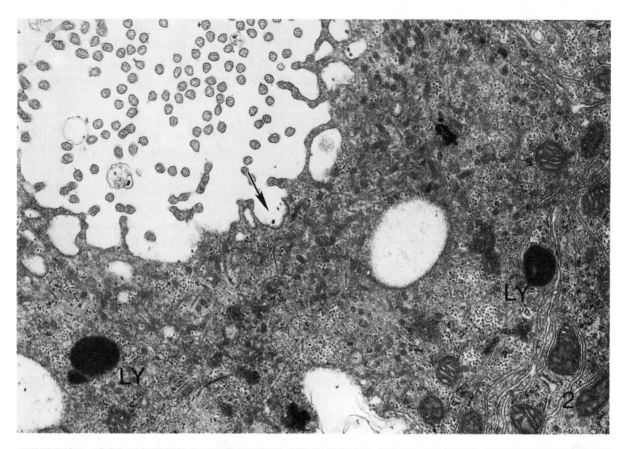

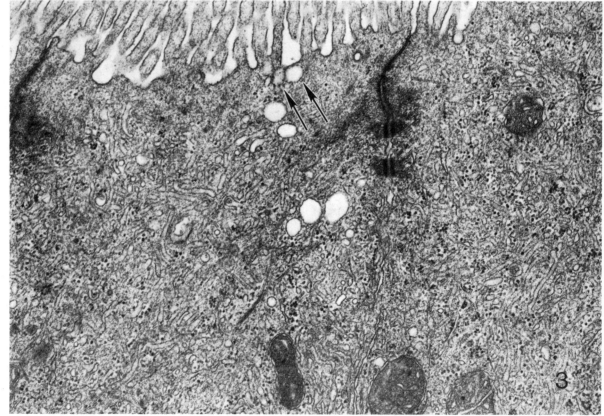

42

EFFECT OF HYPOTONICITY ON CHLORIDE TRANSPORT IN THE ISOLATED OPERCULAR EPITHELIUM OF FUNDULUS HETEROCLITUS

Jose A. Zadunaisky and Lisa Au

Department of Physiology and Biophysics, New York
University Medical Center, New York, N.Y. 10016

The rapid signal detected by <u>Fundulus h</u>. during acclimation to media of diverse salinity is apparently the total osmolarity of the plasma (Zadunaisky J.A. et al. 1992, Bull.MDIBL,31,152-156). For the transition from fresh water to sea water we concluded that the NaK2Cl cotransporter, and the Na/H exchanger were involved in this acclimation that occurs through cell volume regulation. Now we have examined the effects of hypotonicity on the isolated opercular epithelium. A reduction in osmolarity of the solutions bathing the isolated preparation was used as an experimental equivalent of the hypotonicity encountered when the fish moves from sea water to fresh water.

The isolated opercular preparation (see Zadunaisky, 1984, Fish Physiology, Vol. Xb) was mounted, voltage and short circuit current measured and after preliminary tests, the osmolarity of the solution bathing the basolateral side was reduced by replacement with dilute salt solution or plain distilled water. The short circuit current was reduced when the osmolarity was decreased in the basolateral side, no effect was found when it was reduced in the apical side, of these preparations obtained from sea water adapted <u>Fundulus</u>. Figure 1 shows the reduction in current for an average of 6 points for each of 6 osmolarities below the initial one of the isotonic solution. The values in Fig. 1, 6.3,12.5,25,50 and 100 indicate the reduction in miliosomoles produced on the basolateral side. It can be observed that a small reduction in osmolarity can produce a substantial change in chloride secretion.

Curve of Osmolarities

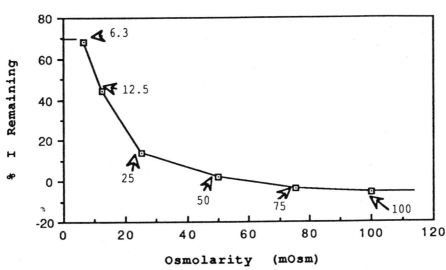

Fig. 1. Reduction in I (short circuit) at different hypotonic levels in the basolateral side of opercular epithelia. Each point is the mean of six experiments.

In order to understand which mechanisms are involved in the response to hypotonicity, the effect of 25 mOsm reduction in basolateral osmolarity was used as a standard test, while the preparation was treated with specific inhibitors of mechanisms existing in the cell membranes of the chloride cells. Thus it was found that 10^{-4} M DIDS partially inhibited the drop in chloride current produced by hypotonicity. In 5 experiments the controls dropped after 25 mOsm of hypotonicity to 13.7 % of the initial current while after DIDS they only dropped to 46.1. This implies that the Cl/HCO$_3$ exchanger of the basolateral membrane is of importance in the adaptation to lower salinities. In contrast the inhibition of the Na/H exchanger with amiloride 10^{-3}M and of the NaK2Cl cotransporter with furosemide 10^{-4}M did not modify significantly the response to hypotonicity. Because of the importance of K channels for cell volume regulation, we tested also quinine at a concentration 10^{-4}M on the preparation, alone and in combination with the effects of hypotonicity. We found that, in effect, quinine tended to reduce the current drop produced by hypotonicity indicating that part of the response involves the opening of a K channel. No effect was found of quinine on the response to mannitol in hypertonic concentrations, and we have to conclude that the activation of a K channel is involved only in the hypotonic response.

These results indicate that volume regulation with a vectorial response in chloride secretion occurs not only during adaptation to higher salinities but also during exposure to lower ones.

Acknowledgment: This research was supported by NIH grant EY1340
to JAZ.

PRESENCE OF NON-HEMATOPOIETIC CARBONIC ANHYDRASE IN THE KIDNEYS OF TWO SALTWATER TELEOSTS, PSEUDOPLEURONECTES AMERICANUS AND FUNDULUS HETEROCLITUS

Erik R. Swenson[1], Brian C. Taschner[2],
David S. Miller[3], Thomas H. Maren[2], and J. Larry Renfro[4]

[1]Department of Medicine, University of Washington, Seattle, WA, 98108
[2]Department of Pharmacology and Therapeutics, University of Florida, Gainesville, FL 32610
[3]Laboratory of Cellular and Molecular Pharmacology, NIH-NIEHS, Research Triangle Park, NC, 27709
[4]Department of Physiology and Neurobiology, University of Connecticut, Storrs, CT, 06269

Our previous physiological and pharmacological work (Swenson et al., Bull. MDIBL 14:127, 1974, Deetjen and Maren, Pflugers Arch 346:25,1974; Swenson and Maren, Am J Physiol 250:F288,1986; Maren et al. Am J Physiol 263:F49,1992) has upheld the paradigm advanced by Homer Smith that seagoing fish (both stenohaline and salt water adapted euryhaline species) lack renal carbonic anhydrase (CA). Freshwater fish and euryhaline freshwater adapted species have readily detectable renal CA activity and function. Attempts, however, from the start to confirm biochemically that seawater fish have no renal CA have been frustrated by the presence of CA in hematopoietic cells intimately associated with the renal tubules (Hodler et al. Am J Physiol 183:155,1955). Our many efforts in the euryhaline eel and salmon, sculpin, and elasmobranch shark and skate to tease apart tubules from the renal matrix, subtract non-renal CA content indirectly by hemoglobin analysis or extract renal CA mRNA have been difficult and unproductive (Gehnrich et al. Bull MDIBL 33:61,1994; Gehnrich et al., Bull MDIBL 34:96,1995 and 34:83, 1995). The winter flounder, Pseudopleuronectes americanus, although not naturally euryhaline, is able to acclimatize to near zero salinity and has renal tubules which are easily dissected free from surrounding hematopoietic tissue for direct analysis and in vitro culture.   These characteristics encouraged us to investigate the problem pharmacologically in vivo, and in vitro with CA activity measurements of isolated tubules and cultured renal epithelium. Additionally, we performed immunofluorescence studies on isolated tubules from saltwater adapted flounder and both the salt and freshwater adapted euryhaline killifish, Fundulus heteroclitus, with antibodies to mammalian CA.

Flounder and killifish of both sexes were used in these studies.   The fish were anesthetized with MS-222 (0.5 g/l seawater) for either sacrifice to obtain kidney tissue or placement of a PE 190 catheter in the urinary papilla of the flounder for urine excretion studies.   Preparation of renal tubules from the flounder was as previously reported (Dickman and Renfro, AJP 251:F424, 1986; Gupta and Renfro, AJP 256:R850, 1989). After the animals were exsanguinated and decapitated, residual blood was removed by perfusing the dorsal aorta with ice-cold solution of M-199 (Sigma Chem. Co).   Kidneys were excised, weighed, and teased apart.   Adherent extrarenal tissue was removed by suspension in 45 ml Ca and Mg- free saline with 0.2% trypsin for 45 min at 22 $^\circ$ C followed by trituration with a 10 ml pipette.   Dissociated extrarenal cells were then separated from the tubules by filtration through Nitex (20 um). Flounder renal epithelial cells were released from tubule fragments by 3 days of trypsinization at 5 $^\circ$ C. Released cells were collected through 50 um Nitex mesh, washed, pelleted, suspended in culture medium, and plated to confluency on native rat tail collagen (Dickman and Renfro, Soc Exp Biol Sem Series 52:65, 1993). After 4 days, collagen gels were released from the plastic

45

35 mm plates and allowed to float as collagen rafts. After 12 days, the collagen gels had been contracted by the confluent monolayers from the initial 35 mm diameter to 17 mm.

Carbonic anhydrase activity in dissected tubules and cultured cells was measured using our micromethod (Maren, J Pharmacol Expt Therap 130:389, 1960). Immuno-fluorescence studies using fluorescent tagged antibodies to mammalian CA II and IV isozymes were performed on tubules from salt water adapted flounder and killifish and in freshwater adapted killifish. Tubule suspensions were fixed, permeabilized, stained and then placed on glass cover slips. The tubules were viewed with an inverted microscope equipped with epi-fluorescence optics and a video camera linked to a MacIntosh computer (Miller and Pritchard Am J Physiol 267:R695, 1994).

After a 24 four hour recovery from anesthesia and urinary catheter placement, urine output was collected over 4 hours. The fish was then given 200 mg/kg of methazolamide by intraperitoneal injection. Urine was collected for 4 hours. Urinary pH was measured with a calibrated pH meter and total CO_2 manometrically with a Kopp-Natelson gasometer.

CA activities in both freshly dissected tubules and isolated cultured cells were similar with 200-300 and 500-600 enzyme units per ml of cells respectively. CA activity in the hematopoietic supernatant of the tubule preparation was equal to that in the isolated tubules; ~200-300 enzyme units per ml. The response of the salt water adapted flounder to methazolamide was negative. There was no significant change in urinary volume, pH or total CO_2, exactly as we have shown in the seawater adapted eel, the teleost sculpin and elasmobranch shark and skate. However, immunofluorescence studies revealed an abundant specific binding of CA II and IV antibodies to peritubular myoepithelial cells surrounding the tubules but none in the proximal tubular epithelium of the salt water adapted flounder and killifish (Figure 1). In contrast, freshwater adapted killifish showed specific fluorescence in the proximal tubular epithelial cytosol with CA II antibodies.

These results taken as a whole support the concept that seagoing fish lack renal epithelial carbonic anhydrase and that it is only in freshwater that fish express renal CA biochemically and functionally in urinary acidification and bicarbonate reabsorption. The immunofluorescence studies, however, demonstrate that salt water fish have an antigenically detectable mammalian CA-like protein in peritubular adventitial cells but lack it in the tubular epithelium. Its function is unknown. If both isolated tubules and cultured epithelial cells contain these peritubular cells this may explain the paradoxical finding of CA activity in vitro yet no pharmacological effect of methazolamide on urinary pH and bicarbonate absorption in the whole animal. However, the picture is not entirely clear for two reasons. First, the filtered load of bicarbonate in seawater fish (as opposed to freshwater fish) is very low and may be accomplished by other non CA-dependent mechanisms of acidification and bicarbonate reabsorption (Deetjen and Maren, Pflugers Arch 346:25,1974; and Swenson et al. AJP 267:F639, 1994). Second, acetazolamide reduces sulfate transport in cultured epithelial cells of the salt water adapted flounder (Renfro, unpublished data). Thus, it is possible that a small amount of CA (not detected with mammalian antibodies) subserves other transepithelial transport processes in the seagoing fish and only becomes critical for the higher rates of urinary acidification and bicarbonate reabsorption in freshwater. Further work in flounder isolated tubules and cultured renal epithelium may permit us to generate specific antibodies to fish renal CA and isolate renal CA mRNA for future studies in the expression, function, and control of renal carbonic anhydrase.

Supported by NIH grant HL-45571(ERS), Univ. of Florida sponsored research grant (THM), and NIH-NIEHS (DSM).

EFFECT OF MEMBRANE-BOUND CARBONIC ANHYDRASE (CA) INHIBITION ON BICARBONATE EXCRETION IN THE SHARK, SQUALUS ACANTHIAS

Erik R. Swenson[1], Brian C. Taschner[2], and Thomas H. Maren[2]

[1]Department of Medicine, VA Medical Center and University of Washington, Seattle, WA, 98108
[2]Department of Pharmacology and Therapeutics, University of Florida, Gainesville, FL 32610

We reported last year (Swenson et al., Bull. MDIBL. 34:94,1995) on the effects in the elasmobranch shark of a high molecular weight polymer linked to a carbonic anhydrase inhibitor; polyoxyethylene-aminobenzolamide. Its high molecular weight (3700) and water solubility limit its distribution to extracellular space and restricts its inhibition to CA on cell surfaces. Indeed, we found no uptake by red cells, gill or muscle of the shark. An intravenous dose of 50 mg/kg had no effect on arterial PO_2, pH or PCO_2 in the normal fish but slowed the rate of gill bicarbonate excretion following an intravenous load of $NaHCO_3$. However, the effect was less than that of benzolamide, whose uptake into gill inhibits both intracellular and plasma membrane-bound CA. These results were interpreted as either submaximal inhibition of gill surface membrane-bound CA or independent additive roles for CA isozymes in gill HCO_3^- excretion. To distinguish between these possibilities a dose response study was undertaken. Squalus acanthias (wt. range 1.8-2.2 kg) were studied 12-16 hr after transfer into small Plexiglas tanks and placement of a dorsal artery catheter. A metabolic alkalosis was induced by a 1 hr infusion of 1 M $NaHCO_3$ (9 mmol/kg). At the start of the $NaHCO_3$ infusion 25, 100 or 200 mg/kg of polymer-linked inhibitor was given intravenously over 5 min. Arterial blood was sampled hourly for pH, total CO_2 and PO_2.

The table shows the time course of plasma HCO_3^- in mM with polymer inhibitor and compares it to the rapid normal (control) rate and to the suppressed rate of normalization by total gill CA inhibition with benzolamide (Swenson and Maren, Am J Physiol 253:R450,1987). Despite a four fold increase in the dose, there was no further statistically significant slowing in the rate of bicarbonate clearance. The effect of the lower dose of 25 mg/kg did not differ from the higher doses. These results suggest that bicarbonate excretion by the gill requires CA activity at both the plasma membrane and in the cytosol since the polymer inhibitor clearly causes HCO_3^- retention but less than benzolamide. In this regard, the gill appears similar to the mammalian kidney. Several studies involving CA II deficient mice and humans as well as selective inhibition of CA IV point to an additive role for both intracellular CA II and brush border membrane-bound CA IV activity in normal proximal tubular HCO_3^- reabsorption (Brechue, Kinne-Saffran, Kinne, and Maren. Biochim Biophys Acta 1066:201,1991; Sly, Whyte, Krupin, Sundaram. Pediatr Res 19:1033, 1985; and Hsu, Moeckel, and Lai. J Am Soc Nephrol 6:701A,1995).

Time hr	Control	Benzolamide 2 mg/kg	Polymer-linked inhibitor			
			25 mg/kg	50 mg/kg	100 mg/kg	200 mg/kg
			plasma HCO_3^- (mM)			
0	4.8 (0.3)	5.1 (0.3)	5.1 (0.3)	5.0 (0.4)	4.9 (0.3)	5.2 (0.5)
1	35 (1.5)	37 (1.4)	33 (1.4)	37 (1.5)	38 (1.7)	39 (1.6)
2	14 (0.9)	29 (0.8)	22 (1.0)*	24 (1.4)*	22 (1.3)*	26 (1.5)*
3	10 (0.6)	24 (0.5)	19 (1.0)*	17 (0.8)*	17 (0.7)*	21 (1.0)*
4	8 (0.5)	21 (0.6)	16 (0.6)*	15 (0.7)*	12 (0.4)*	16 (0.8)*
n =	(8)	(7)	(5)	(5)	(6)	(3)

values are means $\pm$ (SD), * $p < 0.05$ vs. control and benzolamide

Supported by NIH grant HL-45571(ERS) and Univ. of Florida sponsored research grant (THM).

EVIDENCE FOR AN AMILORIDE SENSITIVE GILL Na+/H+ EXCHANGE DURING THE RECOVERY FROM ACIDOSIS IN THE LONG-HORNED SCULPIN (<u>MYOXOCEPHALUS OCTODECIMSPINOSUS</u>)

James B. Claiborne, Jennifer Campbell and Layron Long
Department of Biology, Georgia Southern University, Statesboro, GA 30460

Acid-base transfers in the long-horned sculpin are altered when ambient [Na+] is reduced (Claiborne, Walton & Compton-McCullough, J. Exp. Biol. 193:79-95, 1994; Claiborne, Perry & Bellows, Bull. MDIBL 32:95-97, 1993). Transbranchial Na+/H+ exchange may assist the animal in compensating for internal acidosis, and an external [Na+] of 20-30 mmol l^{-1} may be required (Claiborne & Bellows, Bull. MDIBL 34:63, 1995). In contrast, some studies on freshwater trout have indicated that excretion of H+ is accomplished by an electrogenic H+ ATPase linked to the 1:1 uptake of Na+ through apical Na+ channels (e.g., Lin and Randall, J. Exp. Biol. 161:119-134, 1991). These authors found that H+ exchange across the gills showed little amiloride sensitivity. Thus, in the present study, we have tested the effect of amiloride and a more specific analog of amiloride (5-N,N-hexamethylene-amiloride; a specific inhibitor of the Na+/H+ antiport; Kleyman and Cragoe, J. Mem. Biol. 105:1, 1988) on the net H+ excretion from the sculpin following acid infusion.

All animals were pre-adapted to dilute seawater (20%; [Cl-]: ~100 mmol l^{-1}) for 9-10 days. Fish were then fitted with an intraperitoneal cannula and following an overnight control period, infused with dilute HCl (2.0 mmol kg^{-1}). External water samples were collected periodically throughout the experiment and analyzed for net transfers of H+ (ΔH+) according to the methods of Claiborne et al. (1994). Following the infusion, ΔH+ was measured during a three hour post-infusion period and then a second three hour period in which amiloride (or hexamethylene-amiloride; HMA) had been added to the water (to a final concentration of 1 x 10^{-4} M). Finally, the ambient water was flushed and ΔH+ was determined during two recovery periods (1.5 and 5.5 hours in length). Dunnett's tests were used for comparisons of ΔH+ between control and experimental periods after using the Bonferroni procedure to control for error rate in repeated-measures data (overall protection level, 0.05).

As shown in Figure 1, ΔH+ was reduced to values not significantly different from control levels following the acid infusion when either form of amiloride was added to the external water. The effect was reversible as ΔH+ increased once again during the first recovery period when the water was returned to normal. By the end of the second recovery period, approximately 160 and 200% (3.2 and 4.0 mmol kg^{-1}, respectively for the amiloride and HMA series) of the infused load had been excreted by the fish. The "over-excretion" of acid during the recovery periods is not surprising as we have

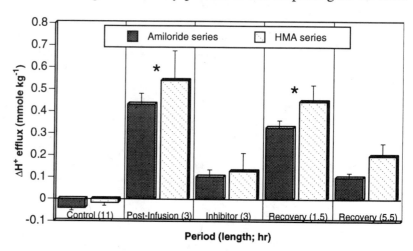

Figure 1. H+ transfers following acid infusion and the addition of amiloride or hexamethylene-amiloride (HMA) to the external water. A positive transfer rate indicates a net efflux from the animal. An "*" indicates ΔH+ rates that were significantly different from the control period (mean ± S.E., p<0.05, n=6 for amiloride and n=4 for HMA series).

48

previously hypothesized that once gill H^+ transfer mechanisms are activated, they remain functional after the initial acidosis has been compensated (Claiborne and Perry, Bull. MDIBL 31:54-56, 1992).

Thus, in contrast to some data for freshwater species (Lin and Randall, 1991) amiloride does inhibit H^+ transfers from the sculpin during acidosis. While it is possible that amiloride indirectly reduced electrogenic H^+ transport by blocking Na^+ channels, ΔH^+ inhibition of a similar magnitude was also demonstrated using the HMA analog. As this analog is thought to be a specific inhibitor of the Na^+/H^+ antiporter in mammalian systems, these results point to the presence of a coupled Na^+/H^+ exchange in the sculpin gill. These findings also agree with preliminary Northern blot studies which detected the presence of mRNA transcripts in the sculpin which were homologous to a human cDNA probe for the NHE-1 isoform of the Na^+/H^+ exchanger (Harris, Claiborne, Pouyssegur and Dawson, Bull. MDIBL 32:128-130, 1993). It remains to be seen if regulatory changes of this transporter at the molecular level can be linked to observed physiological adjustments measured in vivo.

This study was funded by NSF RUI 94-19849 to J.B.C. and NSF REU 93-22221 to J.C.

AMMONIA DISTRIBUTION BETWEEN BLOOD AND TISSUES
IN LATE-TERM EMBRYOS OF THE DOGFISH, SQUALUS ACANTHIAS

Gregg A. Kormanik[1] and Cassandra Harris[2]
[1]Dept. of Biology, Univ. of North Carolina at Asheville, NC 28804
[2]Webster University, St. Louis, MO 63114

Ammonia is a highly reduced form of nitrogen excreted by many aquatic fish (Wood, in Fish Physiology, CRC Press, ed. D.H. Evans, pp. 379-425, 1993). Ammonia occurs in two forms: as NH_3, which behaves as a respiratory gas, and as NH_4^+, a charged ion. The relative concentration of each depends on the pH. Since the pK'_a of NH_4^+ is about 9.25, at physiological and environmental aquatic pH (ca. 7.5-8.0), the NH_4^+ form predominates. The distribution of ammonia across membranes depends on the membrane permeability to each form, and the electrical and chemical gradients. Wood (1993) reviews the factors that affect ammonia distribution. If $P_{NH3} >> P_{NH4+}$, then the pH gradient determines the distribution (e.g. mammals, where the ratio of intra- to extracellular ammonia is about 3). If P_{NH4+} is "significant" (see below), then the transmembrane potential dominates the steady-state distribution (Wood, 1993). Data for teleost fish suggest a much higher ratio for intra- to extracellular ammonia (e.g. approaching 35, Wright et al., J. Exp. Biol. 136, 149, 1988a; Wright et al., J. Exp. Biol. 134, 123, 1988b; Wright and Wood, Fish Physiol. Biochem. 5, 159, 1988), and a trend toward the predominance of P_{NH3} during the evolution of fish to amphibians to mammals (Wood, 1993). However, in shark gill epithelium, previous data indicate that NH_3 permeability predominates, since $P_{NH3}:P_{NH4+}$ is 1000:1 (Evans and More, J. Exp. Biol. 138, 375, 1988). We examined the distribution of ammonia between blood and tissues in late-term dogfish embryos, to determine whether dogfish more closely resemble the teleost or mammalian model.

Embryos were collected from pregnant dogfish (Squalus acanthias) as previously described (Kormanik and Evans, J. Exp. Biol. 125:173-179, 1986). Fish were acclimated to fresh seawater (15° C.), or were exposed to seawater with ammonia concentrations adjusted from 0 to 3 mM (with NH_4Cl), about the highest they could tolerate (3 of 5 died). Fish were acclimated for 24 to 48 hours. Blood ammonia reached maximum levels and stabilized by 24 hours (not shown). At the end of the experimental periods, animals were killed and blood and tissue samples taken and processed as previously described (Kormanik and Verity, Bull. MDIBL 34:92-93, 1995). Ammonia was determined on the deproteinized extracts of tissues and plasma using an enzymatic assay (Sigma 170-UV). Water content of the tissues was determined by drying samples at 70° C. to constant weight (ca. 24 hrs). Tissue ammonia concentrations were calculated using ICF Tamm = (wet Tamm - ECFV * plasma [Tamm])/ICFV, where I- and E- refer to intra- and extracellular and FV refers to fluid volume. ECFV was taken as 12.7% (Robertson, Biol. Bull. 148, 303. 1975), and ICFV = Water content - ECFV. Tamm refers to total ammonia (NH_3 + NH_4^+).

The results of the experiment are presented in Figure 1, where tissue Tamm is plotted against plasma Tamm for liver, muscle and brain. Relevant data for these curves are included in Table 1. Liver tissue exhibited the greatest slope, and the highest concentrations of Tamm. Liver tissue also had the lowest water content (24.2 ± 0.98 %, n=5) compared to brain (79.7 ± 0.88 %, n=3), muscle (70.5 ± 0.71 %, n=5) and plasma (88.8 ± 0.32 %, n=3). The latter were comparable to data for adult fish. Liver water content reported for these embryos is low, which may be attributable to the high lipid content of liver in this lecithotrophic species. Slopes ($Tamm_i/Tamm_e$) for these tissues ranged from 4.01 to 7.33. All slopes were highly correlated (p < 0.01).

Roos and Boron (1981; in Wood, 1993) present an equation which describes the steady-state distribution of Tamm between intra- and extracellular compartments, relating $Tamm_i/Tamm_e$ to the transmembrane potential, pH, electrochemical gradient and ratio of the permeabilities of NH_3 and NH_4^+. This equation was used to calculate P_{NH3}/P_{NH4+} ratios presented in Table 1. Assumptions included pH_e = 7.9, pH_i = 7.4, transmembrane voltage = -90mV, T = 15° C., and pK'_a = 9.25, which are

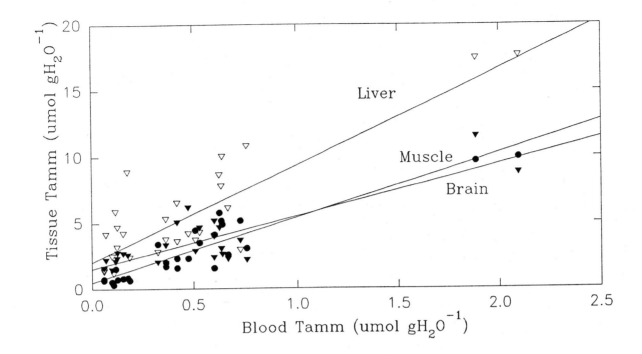

Figure 1. Tissue total ammonia (Tamm) versus plasma total ammonia (Tamm) in liver, muscle and brain of <u>Squalus</u> <u>acanthias</u> embryos acclimated to elevated ambient ammonia concentrations. Open triangles, liver; closed triangles, muscle; closed circles, brain.

Table 1. Relationship between $Tamm_i/Tamm_e$ and the permeability of NH_3 and NH_4^+. Correlation coefficients (r) and significance (p) are presented for the data in Figure 1.

Tissue	$Tamm_i/Tamm_e$	P_{NH3}/P_{NH4+}	n (pairs)	r	p <
Brain	4.01	242	27	0.738	0.01
Muscle	4.90	121	30	0.923	0.01
Liver	7.33	48	30	0.846	0.01

reasonable values for these fish. When P_{NH3}/P_{NH4+} approaches 0.1, $Tamm_i/Tamm_e$ asymptotically approaches 35. When P_{NH3}/P_{NH4+} approaches 100, $Tamm_i/Tamm_e$ asymptotically approaches 3 (Wood, 1993). In elasmobranch brain, muscle and liver, P_{NH3}/P_{NH4+} ranges from 242 to 48, indicating that the permeability of NH_3 predominates in the distribution of ammonia in these tissues.

These data contrast with those of Wright et al. 1988a; 1988b and Wright and Wood, 1988) who report $Tamm_i/Tamm_e$ ratios of 30-35 in white muscle at rest for lemon sole and trout. These $Tamm_i/Tamm_e$ ratios correspond to P_{NH3}/P_{NH4+} ratios of 1.7-1. Thus for Tamm to be distributed according to transmembrane potential, rather than to transmembrane pH, P_{NH4+} must be "significant", but not necessarily larger than P_{NH3} (Wood, 1993). Our data, however, indicate that the permeability to NH_3 predominates in these elasmobranch tissues, and dogfish resemble the mammalian model, with ammonia distributed according to transmembrane pH. Liver had the highest $Tamm_i/Tamm_e$ ratio, indicating a relatively higher permeability to NH_4^+. This higher Tamm content in liver tissue may be indicative of ammonia "trapping", resulting from a lower P_{NH3}/P_{NH4+} ratio (Table 1) that would facilitate movement of blood ammonia nitrogen into the liver for urea synthesis. (Supported by NSF IBN-9507456 to GAK; NSF REU 9322221 for CH).

QUINIDINE INHIBITS TAURINE TRANSPORT BY THE COELOMOCYTES
OF THE MARINE POLYCHAETE, GLYCERA DIBRANCHIATA

Robert L. Preston[1], Marianne T. Kaleta[1], Keith M. Katsma[1],
Graciana Lapetina[2], Kristin A. Simokat[3] and Paula R. Zimmermann[1]
[1]Department of Biological Sciences
Illinois State University, Normal, IL. 61790-4120
[2]Princeton University, Princeton, NJ 08544
[3]Wesleyan University, Middletown, CT 06459

Earlier investigations in our laboratory have shown that taurine transport by the hemoglobin containing coelomocytes (red blood cells, RBCs) of the marine polychaete, Glycera dibranchiata, is rapidly inhibited by exposure to micromolar concentrations of mercuric chloride (Chen, C.W. and Preston, R. L., Bull Environ. Contam. Toxicol. 39:202-208, 1987; Preston, R. L. and Chen, C.W., Bull Environ. Contam. Toxicol. 42:620-627, 1989). It is probable that mercuric chloride reacts with sulfhydryl groups associated with the membrane transport carrier for taurine since under the conditions employed in these experiments there are no apparent changes in ion gradients or membrane potential that might influence taurine transport indirectly (Preston, R. L. and Chen, C.W., Bull Environ. Contam. Toxicol. 42:620-627, 1989; Preston, R. L., Truong, T. T., Lu, S. and Janssen, S. J., Bull. MDIBL 29:78-81, 1990; Wondergem, personal communication). Recent experiments have also shown that the reactive form of mercury is most likely the $HgCl_3^-$ complex (Preston, R. L., Zimmermann, P. R., Kaleta, M. T. and Simokat, K. A., Bull. MDIBL 33:53-55, 1994). The fact that the anionic form of mercury is most effective suggests that inhibitors of anion channels might also have some effect on taurine transport. Quinine and its analogue, quinidine, have been shown to inhibit anion channels involved in volume regulation (Banderali, U. and Roy, G. J. Membr. Biol. 126:219-234, 1992; Sanchez Olea, R. Pasantes-Morales, H. Lazaro, A. and Cereijido, M., J. Membr. Biol. 121:1-9, 1991). In this series of experiments, we show that quinidine rapidly and irreversibly inhibits taurine transport by Glycera RBCs.

Glycera RBCs were removed from the animals, washed in artificial seawater (NaSW) and separated from gametes and other coelomocytes by differential centrifugation. All experiments were done at 12°C. Quinidine (0.01 mM - 5 mM) was dissolved in NaSW which was then preincubated with Glycera RBCs for 5 min. Controls were incubated in medium containing only NaSW. Taurine influx was measured by incubating the RBCs with 1 mM ^{14}C-taurine in NaSW for 5 minutes. The RBCs were then separated from the radioactive medium by centrifuging the cells through dibutylphthalate (Chen, C.W. and Preston, R. L., Bull Environ. Contam. Toxicol. 39:202-208, 1987). Trichloroacetic acid extracts of the RBCs were transferred to scintillation vials and isotope content determined by scintillation spectroscopy. The data were corrected for cell number by measuring hemoglobin content with Drabkin's reagent which is directly correlated with cell number and cell water content. In some experiments, RBCs were also incubated with 20 μM $HgCl_3^-$ for 1 min and or 1 mM quinidine for 1 min to determine possible additive effects of these agents. All experiments reported here have been repeated in triplicate at a minimum.

Preliminary experiments in which Glycera RBCs were incubated with 1 mM quinidine for times ranging from 10 sec to 5 min showed that the inhibition of taurine influx occurred quickly (< 1 min), reaching a plateau from 1 min through 5 min of about 50% inhibition. Other experiments also showed that the effect of quinidine was not readily reversible with repeated washing. Therefore, in most subsequent experiments the RBCs were treated with quinidine, washed in NaSW and then incubated with ^{14}C-taurine for flux measurement.

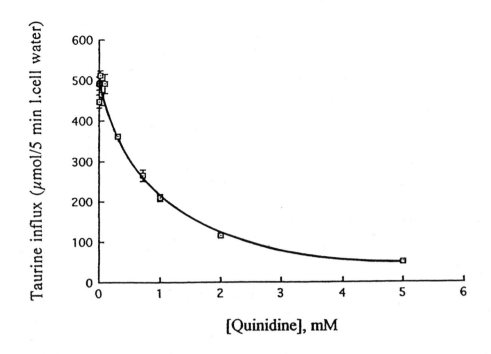

Fig 1. Quinidine Inhibition of Taurine Transport.
   ^{14}C-Taurine transport by <u>Glycera</u> RBCs was measured in NaSW after they
   were preincubated with quinidine at various concentrations for 5 min.

   The concentration dependence of quinidine inhibition is shown in Fig 1.
The estimated half-inhibition concentration is about 0.8 mM. This
concentration is similar to that used by Banderali and Roy (1 mM; J. Membr.
Biol. 126:219-234, 1992) to block Cl[-] channels in MDCK cells. However, they
found that the quinidine can be washed out and the inhibitory effect is
reversible in their system.

   The fact that quinidine inhibits taurine transport and that the anionic
form of mercury, $HgCl_3^-$, is most effective in inhibition of taurine transport
suggests it is possible that one or the other of these compounds could
"protect" or block access of the other to reactive sites on the transporter.
To test this hypothesis, <u>Glycera</u> RBCs were incubated sequentially with 20 μM
mercuric chloride followed by 1 mM quinidine. In another set of samples the
order of incubation was reversed. In addition, controls were included for
untreated cells and for each agent incubated alone with RBCs. Table 1 shows
the results of this experiment.

   If mercuric chloride and quinidine act independently, one would expect
that their inhibitory effects would be additive. This prediction seems to be
borne out in the case of mercury treatment followed by quinidine (Hg/Quin,
Table 1). The inhibition ratio compared with the NaSW control (J_I/J_0) for
quinidine and mercuric chloride alone was 0.56 and 0.30 respectively. The
predicted J_I/J_0 after treatment to both agents should therefore be 0.56 x 0.30
= 0.17. The observed value for J_I/J_0 was 0.18. However, if the order of
exposure was reversed (Quin/Hg) the observed value for J_I/J_0 was 0.28. This
is not significantly different from the inhibition due to mercuric chloride
alone. These data suggest that prior treatment of the RBCs with quinidine may
partially prevent mercurial inhibition of taurine transport. If quinidine

Table 1: Effect of Order of Exposure to Quinidine and Mercuric Chloride on Taurine Transport by Glycera RBCs.

Preincubation	Taurine influx*	(J_I/J_0)**	p
NaSW	1118 $\pm$ 55	–	
Quinidine (1 mM)	626 $\pm$ 34	0.56	<0.001
HgCl$_2$ (20 μM)	342 $\pm$ 23***	0.30	<0.001
Quin/Hg	311 $\pm$ 24	0.28	<0.001
Hg/Quin	206 $\pm$ 9	0.18	<0.001

* μmol. 5 min^{-1} l.cell water^{-1} ($\pm$ S.E., n = 6)

** (J_I/J_0) = inhibition ratio where J_I = taurine influx after exposure to mercury, quinidine or both; J_0 = control taurine influx in NaSW. Quin/Hg = 1 mM quinidine incubation for 1 min followed by 20 μM HgCl$_2$ incubation for 1 min. The order of exposure was reversed in the Hg/Quin condition. Cells were then washed and taurine influx measured.

*** HgCl$_2$ treatment is not significantly different than the Quin/Hg treatment (p > 0.4). All other comparisons are significantly different (p $\leq$ 0.005). Student's t-test was used to compare the statistical significance (p) of the experimental conditions to the control.

completely blocked mercuric chloride inhibition then J_I/J_0 should equal 0.56 as in the quinidine control. The fact that the reactive form of mercury appears to be HgCl$_3^-$ suggests that the ability of quinidine to block anion channels extends to blocking anionic mercury as well. Perhaps this involves blocking access of mercuric chloride to reactive sites directly or allosteric changes in the transport protein following quinidine binding at occludes site reactive with mercury. It is also possible that this interaction occurs by more indirect means since both quinidine and especially mercuric chloride are somewhat nonspecific in their reactivity. For the present, at least, these data are consistent with the notion that the taurine transporter in Glycera RBCs has some characteristics that resemble an anion channel or anion selective transport system. Future investigations will test this hypothesis further.

Supported in part by NIEHS grant P30-ESO3828. Paula Zimmermann was supported by a NOAA/Sea Grant awarded to RLP. Marianne Kaleta was a recipient of an Undergraduate Research Fellowship from the American Heart Association, Maine Affiliate. Keith Katsma and Kristin Simokat were recipients of Grass Fellowships. Graciana Lapetina was a recipient of an NSF-REU Undergraduate Research Fellowship.

EFFECT OF MEDIA ANION COMPOSITION ON MERCURY INHIBITION OF TAURINE TRANSPORT BY THE COELOMOCYTES OF THE MARINE POLYCHAETE, GLYCERA DIBRANCHIATA

Robert L. Preston[1], Keith M. Katsma[1], Graciana Lapetina[2]
and Paula R. Zimmermann[1]
[1]Department of Biological Sciences
Illinois State University, Normal, IL. 61790-4120
[2]Princeton University
Princeton, New Jersey 08544

Previous studies in our laboratory have shown that taurine transport by the hemoglobin containing coelomocytes (red blood cells, RBCs) of the marine polychaete, Glycera dibranchiata, is rapidly inhibited by exposure to micromolar concentrations of mercuric chloride (Chen, C.W. and Preston, R. L., Bull Environ. Contam. Toxicol. 39:202-208, 1987; Preston, R. L. and Chen, C.W., Bull Environ. Contam. Toxicol. 42:620-627, 1989). We have concluded that the probable site of action of mercuric chloride is the membrane transport carrier for taurine. It is also likely that mercuric chloride simultaneously modifies other cellular processes because of its high reactivity with sulfhydryl groups (e.g. glucose transport, Preston, R. L. et al., Bull. MDIBL 30:51-53, 1991). However, our evidence supports the notion that mercuric chloride acts directly on the transport carrier or associated moieties rather than by indirect effects (e.g. Preston, R. L. et al., Bull. MDIBL 29:78-81, 1990; Preston, R. L. and Chen, C.W., Bull Environ. Contam. Toxicol. 42:620-627, 1989).

The ionic state of mercury in solution depends on anion concentration (Webb, J.L. in Enzyme and Metabolic Inhibitors, Academic Press, N.Y., 1966). Mercury can exist in a variety of cationic, neutral or anionic forms depending on medium Cl^- concentration.

$$Hg^{++} \leftrightarrow HgCl^+ \leftrightarrow HgCl_2 \leftrightarrow HgCl_3^- \leftrightarrow HgCl_4^=$$

Increasing $[Cl^-] ====>$

In anion substitution studies in which we replaced Cl^- with gluconate, we showed that it is likely that the reactive form of mercury in our system is $HgCl_3^-$ (Preston, R. L. et al., Bull. MDIBL 33: 53-55, 1994). The chloride concentration (approx. 100 mM Cl^-) at which the $HgCl_3^-$ form is maximized correlates well with the concentration at which the inhibition of taurine transport is maximum. In the present set of experiments, we utilized other anion substitutes as well as gluconate to more rigorously test the hypothesis that $HgCl_3^-$ is the critical reactive form of mercury. We also noticed in preliminary studies that the usual pattern of mercury inhibition observed in gluconate and other media (maximum inhibition at 100 mM Cl^-) was not found in bromide and iodide media. Our present data will show that this anomalous behavior may be due to formation of less reactive mercury complexes in these media.

The concentration of Cl^- in incubation medium containing 20 μM mercuric chloride was varied by iso-osmotic replacement of NaCl with the Na salts of the following anions: gluconate, sulfamate, sulfate, nitrate, methylsulfate, isethionate, bromide, iodide and thiocyanate. In the case of sulfate, the medium contained D-mannitol as well to bring the solution to the correct osmotic prssure. Glycera RBCs were washed 2 times in artificial seawater (NaSW), washed 2 times in the appropriate anion substituted medium (without mercury) and then incubated in the mercury containing medium for 1 minute. This medium was then removed, the cells washed once in the appropriate anion substituted medium without mercury. In the controls, all conditions were identical except that the 1 minute incubation was done in mercury free medium.

Taurine influx was measured by incubating the RBCs at 12°C with 1 mM ^{14}C-taurine in NaSW for 5 minutes. The RBCs were then separated from the radioactive medium by centrifuging the cells through dibutylphthalate (Chen, C.W. and Preston, R. L., Bull Environ. Contam. Toxicol. 39:202-208, 1987). Trichloroacetic acid extracts of the RBCs were transferred to scintillation vials and isotope content determined by scintillation spectroscopy. The data were corrected for cell number by measuring hemoglobin content with Drabkin's reagent (Sigma Chemical Co., St. Louis) which is directly correlated with cell number and cell water content. Medium identified as 0 mM Cl$^-$ medium in this study refers to medium in which no Cl$^-$ salts were added. It should be recognized that low levels of contaminating Cl$^-$ is probably present in the medium and cells suspensions (probably <1 mM).

Table 1: Effect of Various Anion Substituted Media on Mercury Inhibition of Taurine Transport.

Taurine influx, μmol. 5 min^{-1} l.cell water^{-1} ($\pm$ S.E., n = 3)*
Incubated with 20μM Hg for 1 min

	**Control	0 Cl$^-$	(J_I/J_0)	100 Cl$^-$	(J_I/J_0)	514 Cl$^-$	(J_I/J_0)
Gluconate	1233 + 71	1206 + 71	0.98	168 + 9	0.14	362 + 52	0.29
Isethionate	1174 + 26	1380 + 50	1.18	276 + 13	0.24	415 + 18	0.35
Methysulfate	1094 + 47	1104 + 18	1.01	173 + 17	0.16	339 + 8	0.31
Nitrate	94 + 46	696 + 60	0.78	117 + 9	0.13	376 + 50	0.42
Sulfamate	867 + 28	938 + 85	1.08	172 + 28	0.20	418 + 36	0.48
Sulfate	841 + 26	712 + 28	0.85	96 + 13	0.11	294 + 17	0.35
Bromide	974 + 26	884 + 29	0.91	783 + 36	0.80	412 + 31	0.42
Iodide	366 + 22	437 + 10	1.19	443 + 19	1.21	415 + 44	1.13

* Data from two separate experiments were combined in this table.
** Control fluxes were determined on cells washed in the appropriate anion substituted medium (0 mM Cl$^-$) but were not exposed to mercuric chloride. Na salts of the anion listed were substituted iso-osmotically for NaCl. D-Mannitol was added to sulfate medium to adjust the osmotic pressure. The Cl$^-$ concentrations are mM. (J_I/J_0) = inhibition ratio where J_I = taurine influx after exposure to mercury; J_0 = control taurine influx.

Table 1 shows the results for inhibition of taurine transport by Glycera RBCs after 1 min incubation with various anion substituted media containing 20 μM mercuric chloride. The results for gluconate are typical: Little or no inhibition occurred in 0 Cl$^-$ medium compared with the control which was not exposed to mercury (J_I/J_0 = 0.98; where J_I = taurine influx after exposure to mercury and J_0 = control taurine influx). At 100 mm Cl$^-$ mercury inhibited taurine influx $\geq$85% (J_I/J_0 = 0.14). In 514 mM Cl$^-$ (the normal NaSW concentration) mercury inhibited taurine influx by $\geq$70% (J_I/J_0 = 0.29). This pattern reflects the shift from the cationic mercury forms to the anionic forms as Cl$^-$ concentration in the medium increases and is consistent with the hypothesis that HgCl$_3^-$, which is in maximum relative concentration at about 100 mM Cl$^-$, is the form that reacts with the taurine transporter. Although 3 Cl$^-$ concentrations were used in this study for screening purposes, more detailed studies using intermediate Cl$^-$ concentrations are entirely consistent with this hypothesis. A similar pattern of inhibition is seen for isethionate, methylsulfate, nitrate, sulfamate and sulfate media. For these media as a group, the inhibition ratios (J_I/J_0) for 0 mM Cl$^-$ ranged from J_I/J_0 = 0.78 for nitrate to J_I/J_0 = 1.18 for isethionate, with most values close to

1.0; for 100 mM Cl⁻ J_I/J_0 = 0.11 for sulfate to J_I/J_0 = 0.24 for isethionate; and for 514 mM Cl⁻ J_I/J_0 = 0.31 for methylsulfate to J_I/J_0 = 0.48 for sulfamate.

In contrast, mercury was substantially less effective in inhibiting taurine transport in 100 mM Cl⁻ medium in which bromide or iodide were used as anion replacements (J_I/J_0 = 0.80 for bromide and J_I/J_0 = 1.21 for iodide). One possible explanation for this may be that both bromide and iodide have higher affinity constants for complexation with mercury and thus would preferentially form $HgBr_n^x$ or HgI_n^x complexes rather than $HgCl_3^-$ (where n may range from 1 to 4 and x from +1 to -2). In addition, one must assume that the bromide and iodide complexes are less permeable to membrane and/or less reactive with taurine transport protein. If this is true, we would predict that adding low concentrations of Br⁻ or I⁻ in the presence of Cl⁻ should lessen the inhibitory effect of mercury. This hypothesis was tested by incubating RBCs with mercury in 100 mM Cl⁻ medium with and without Br⁻ or I⁻ added at concentrations ranging from 0.01 mM to 10 mM (Table 2).

Table 2A,B: Low Concentrations of Bromide and Iodide Reduce Mercury Inhibition of Taurine Transport in 100 mM Chloride Medium.

A:*	Taurine influx,μmol. 5 min⁻¹ l.cell water⁻¹ (± S.E., n = 3)**				
		Incubated with 20μM Hg for 1min			
Control***	0 mM Br⁻	0.01 mM Br⁻	0.1 mM Br⁻	1mM Br⁻	10 mM Br⁻
1250 + 31 ($J_I/\bar{J}_0$)	156 + 5 (0.1̄2)	131 + 7 (0.1̄0)	139 + 15 (0.1̄1)	505 + 25 (0.4̄0)	679 + 53 (0.5̄4)
B:*					
Control***	0 mM I⁻	0.01 mM I⁻	0.1 mM I⁻	1mM I⁻	10 mM I⁻
1250 + 31 ($J_I/\bar{J}_0$)	156 + 5 (0.1̄2)	339 + 20 (0.2̄7)	1268 + 52 (1.0̄1)	1001 + 35 (0.8̄0)	1158 + 22 (0.9̄3)

* The 1 minute incubation of the RBCs with 20 μM mercuric chloride was conducted in 100 mM Cl⁻ medium with Na gluconate replacing the remaining Cl⁻ iso-osmotically. Br⁻ or I⁻ was added in additon to the 100 mM Cl⁻ present in the medium.
** Taurine influx was measured in NaSW which had a Cl⁻ concentration of 514 mM (see methods).
*** Control fluxes were determined on cells washed in 100 mM Cl⁻ medium but were not exposed to mercuric chloride. (J_I/J_0) = inhibition ratio where J_I = taurine influx after exposure to mercury; J_0 = control taurine influx.

The data in Table 2A show that 20 μM mercuric chloride in 100 mM Cl⁻ medium (gluconate replacement) strongly inhibited taurine transport ($J_I/J_0 \cong$ 0.10) in the range of 0 mM - 0.1 mM added Br⁻. However, at 1 mM and 10 mM Br⁻ the effect of mercury was considerably lessened ($J_I/J_0 \cong$ 0.40 and 0.54 respectively). A similar experiment with added I⁻ (Table 2B) shows that the effect of mercury is lessened at the lowest I⁻ concentration tested (0.01 mM, J_I/J_0 = 0.27) compared with the control value of (J_I/J_0 = 0.12) . At higher I⁻

concentrations (0.1 mM to 10 mM), 20 μM mercury had little if any effect (J_I/J_0 = 0.80 to 1.01). These data should also reflect, in a general way, the relative affinities of the halides for mercury in relation to Cl^- and OH^-. As an index of what the relative affinities of halides for mercury may be, one can use the values published by Webb (Webb, J.L. in Enzyme and Metabolic Inhibitors, Academic Press, N.Y., 1966) for dissociation constants for the equilibrium $K_1 = [Hg^{++}][A^-]/[HgA^+]$, (see below, units in parentheses are -log dissociation constants).

$$I^- (12.9) < OH^- (10.3) < Br^- (9.05) < Cl^- (6.74) < gluconate^- (?)$$

The data in Table 2 are generally consistent with this pattern of affinitites. Iodide is effective at concentrations that are, perhaps, 1/100 to 1/1000 that of bromide. Bromide at least partially prevents mercury inhibition at concentrations 1/10 to 1/100 that of Cl^-. The dissociation constant for gluconate was not available in Webb's data but we would predict that gluconate, as well as the other anion substitutes employed in Table 1, would have a substantially lower dissociation constant for mercury complexes. Other explanations for this behavior are possible, but we feel these data are quite consistent with the notion that inactive complexes are formed with Br^- and I^-. If other tissues (intestinal epithelia, for example) resemble Glycera RBCs in general transport characteristics and sensitivity to mercury, one might speculate that low concentrations of bromide or iodide in the diet might substantially reduce the reactivity of inorganic mercury with membrane transporters. This may be another approach to amelioration of mercurial toxicity in some systems.

Supported in part by NIEHS grant P30-ESO3828. Paula Zimmermann was supported by a NOAA/Sea Grant awarded to RLP. Keith Katsma was a recipient of a Grass Fellowship. Graciana Lapetina was a recipient of an NSF Undergraduate Research Fellowship.

ORGANIC ANION UPTAKE BY TRABECULAR MESHWORK (TBM) CELLS FROM HUMAN EYE

David S. Miller[1], David F. Croft[2] and Jose A. Zadunaisky[2]
[1]Laboratory of Cellular and Molecular Pharmacology, NIH-NIEHS, RTP, NC 27709
[2]Department of Physiology and Biophysics, NYU Medical Ctr, NY, NY 10016

Certain fluid compartments of the body, e.g., aqueous humor and cerebrospinal fluid, possess their own "internal kidneys," responsible for the removal of potentially toxic normal metabolites, drugs and drug metabolites (Pritchard and Miller, Physiol. Rev. 73:765, 1993). Many of these substances are transported as charged compounds and the tissues responsible possess separate plasma membrane transport systems for organic anions and organic cations. In the present report, we investigated the organic anion transport properties of cultured trabecular meshwork cells (TBM) from human eye using a fluorescent organic anion, fluorescein (FL), video microscopy and digital image analysis.

Transfected TBM cells from human eye (Alcon Laboratories) were grown at 37° C on 4x4 cm glass cover slips in Dulbecco's Modified Eagle's Medium (DMEM, low glucose) supplemented with 10% fetal bovine serum, glutamine and Pen/Strep. For experiments, cover slips were mounted in a chamber containing DMEM with 1 µM FL and viewed by means of an inverted microscope equipped with epi-fluorescence optics and a video camera and connected to a Macintosh computer (Miller et al., Am. J. Physiol. 264:R882, 1994).

TBM cells incubated in medium with FL rapidly concentrated the dye, reaching steady-state levels within 10 min (Fig. 1A). At steady state, diffuse fluorescence filled both the cytoplasm and nucleus. Also, areas of intense punctate fluorescence were seen throughout the cytoplasm. These were most concentrated in the perinuclear region, which is the thickest part of the cell. FL uptake was abolished when metabolism was inhibited by KCN. Addition of prostaglandin PGE2 (0.5-5 µM) to the medium caused a concentration dependent decrease in FL accumulation (Fig. 1B). At the highest concentration used, PGE2 reduced FL accumulation by 59%. The organic anion, probenecid, at 1 mM, reduced FL uptake by 52%. p-Aminohippurate did not inhibit FL uptake, rather it caused a small increase (30-40%, not shown).

To date organic anion uptake mechanisms have only been found in those cells functioning to absorb or secrete metabolites and xenobiotics, e.g., renal and hepatic epithelial cells. The present preliminary findings indicate that human TBM cells possess a transport system for organic anions that is specific, uphill and dependent on cellular metabolism. This system may play a role in removing drugs and physiologically important metabolites from aqueous humor. Supported by NIH grant EY 01340.

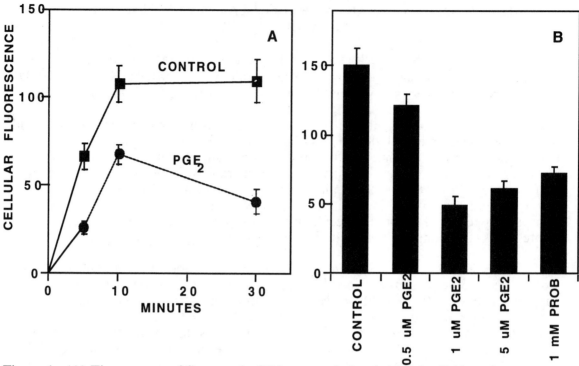

Figure 1. (A) Time course of fluorescein (FL) accumulation in TBM cells from human eye. The PGE_2 concentration was 5 μM. (B) Effects of PGE_2 and probenecid on the 30 min uptake of 1 μM FL. Values are mean pixel intensities $\pm$ SE for 10-19 (A) or 28-29 (B) cells. Statistical comparisons: (A) PGE_2 reduced fluorescence intensity at all times, P<0.01; (B) 1-5 μM PGE2 and probenecid reduced fluorescence intensity, P<0.01.

VOLUME-REGULATORY AMINO ACID TRANSPORT IN HEPATOCYTES FROM <u>RAJA ERINACEA</u>

Ned Ballatori[1], Anh T. Truong[1],
Alistair Donald[2], and James L. Boyer[3]

[1]Department of Environmental Medicine, University of Rochester School of
Medicine, Rochester, NY 14642
[2]Martin Ryan Marine Science Institute, Galway, Ireland
[3]Department of Medicine and Liver Center, Yale University School of Medicine,
New Haven, CT 06510

Many cell types have been shown to regulate their volume following cell swelling by activating a plasma membrane channel that allows taurine and perhaps other intracellular organic osmolytes to efflux from the cell. We demonstrated that skate hepatocytes also possess a swelling-activated osmolyte channel that is permeable to taurine, and have demonstrated that this channel is regulated by intracellular ATP (Ballatori et al., Am. J. Physiol. 267:G285-G291, 1994; Ballatori and Boyer, Am. J. Physiol. 262:G451-G460, 1992; Ballatori et al., Mol. Pharmacol. 48:472-476, 1995).

To further characterize the nature of the channel and its potential substrates, the present study compared volume-activated efflux of ^{14}C-taurine, to that of ^{14}C-L-alanine, ^{14}C-L-phenylalanine, ^{14}C-methylaminoisobutyric acid (MeAIB), ^{14}C-betaine, ^{14}C-glycine, ^{3}H-myoinositol and ^{14}C-sorbitol. Hepatocytes were isolated from male skates and were preloaded with radioisotope by incubating with 0.1 mM of the indicated compounds for 2 h at 15°C. Hepatocytes to be loaded with amino acids were incubated in medium that also contained 2 mM aminooxyacetic acid to inhibit pyridoxal phosphate-dependent enzymes. Cells were then washed to remove extracellular radioactivity, and hypotonicity was induced by diluting the cell suspensions either 40% or 50% with H_2O. Cellular ^{14}C or ^{3}H content at 10, 30 and 60 min after swelling was measured by scintillation spectrometry.

Cell swelling produced a marked activation of ^{14}C-taurine efflux, with ~50% of the amino acid released after one hour of incubation in medium diluted 40% with water. Betaine, glycine, MeAIB, and L-alanine were released at rates comparable to taurine following cell swelling. In contrast, cell swelling produced minimal stimulation of phenylalanine efflux. Comparable findings have previously been reported in skate red blood cells (Haynes and Goldstein, Am. J. Physiol. 265:R173-R179, 1993). However, skate hepatocyte swelling produced only a small increase in myoinositol efflux (10-20% released after one hour) and an even smaller effect on sorbitol efflux (5-10% stimulation of efflux). However, there was considerable spontaneous release of radioisotope from ^{14}C-sorbitol-loaded cells, under isosmotic conditions. It is possible that some of these compounds may have been metabolized by the hepatocytes, although this was not quantitated in the present study.

The present findings indicate that in addition to taurine, other organic osmolytes can be released by skate hepatocytes in response to cell swelling. In particular, small neutral amino acids such as glycine, L-alanine, and betaine are released; however, the larger and more bulky amino acid phenylalanine is not readily released. Sorbitol and myoinositol are also released at a relatively slow rate after cell swelling. Additional studies are needed to distinguish whether these organic osmolytes are released by a single multi-specific channel, or by distinct swelling-activated mechanisms. (Supported by the National Institute of Environmental Health Sciences (ES03828 and ES01247), and the National Institute of Diabetes and Digestive and Kidney Diseases (DK34989 and DK25636)).

THE DEVELOPMENT OF A METHOD FOR MEASURING CELL VOLUME REGULATION IN SINGLE HEPATOCYTES FROM RAJA ERINACEA

C. A. Fletcher[1], A. Donald[2], N. Ballatori[3], M. Nathanson[4] and J. L. Boyer[4]
[1]Westtown School, Westtown, PA 19395-1799;
[2]University College, Galway, Ireland;
[3]Dept of Environmental Medicine, University of Rochester School of Medicine, Rochester, NY 14642;
[4]Dept of Medicine and Liver Center, Yale University School of Medicine, New Haven, CT 06510

All cells in living organisms are constantly exposed to changing conditions in their extracellular environments. Changes in osmolarity cause cells to swell or shrink. In order to maintain homeostasis, cells undergo a process known as regulatory volume decrease (RVD) after cell swelling and regulatory volume increase (RVI) after cell shrinkage. Studies in our laboratory (Ballatori and Boyer, Am. J. Physiol. 262:G451-G460, 1992; Ballatori et al. Am. J. Physiol. 267:G285-G291, 1994; Ballatori et al, Mol. Pharmacol. 48:472-476, 1995) have examined the role of organic osmolytes in regulating the process of RVD. In previous studies volume changes have been measured by isotope dilution methods which provide an average measurement for these changes in large populations of cells in suspension. The present study was designed to measure volume regulatory responses in single cells maintained on cover slips and examined by optical techniques.

To develop this method, skate hepatocytes were isolated as previously described (Smith et al. J. Exp. Zool. 241:291-296, 1987) and allowed to settle on glass coverslips. A second coverslip was suspended above the first by two strips of plastic, then the hepatocytes in the space between the coverslips were perfused with elasmobranch buffer. This perfusion chamber was placed on the stage of a Zeiss IM35 inverted microscope and individual hepatocytes were observed directly. The preparation was maintained at approximately 15°C and cells were visualized using Nomarski optics. Images were captured by a Dage-MTI video camera connected to a Panasonic optical disk recorder. After selecting a field containing several isolated hepatocytes, the cells were subjected to 40% dilution in either Ringer's solution or 40% water and images were recorded at 0.25 minute intervals for one minute during perfusion with normal Ringer's, 0.5 intervals for 5 minutes after diluting with water, then 1 minute intervals for 5 minutes and 2.5 minute intervals for the remaining 20 minutes of the study.

To quantify the volume changes separately for each cell, the images were transferred to a MacIntosh 7100 PC. NIH Image software (version 1.55) was used to trace the outline of individual cells; the measurements were calibrated by measuring beads of known size on the optical disk recorder after calculating the area. The known relationship between area and volume for spheres, $V=(A) \exp 1.5 \times (.7523)$, was used to calculate an average cell volume for each cell.

The figure below illustrates the mean ± SEM for 20 cells normalized by expressing the volumes as a percentage of their control values. Cells swelled immediately after application of hypotonic medium, peaking to approximately 130% of the normal value after 1-2 minutes. Cells then spontaneously underwent regulatory volume decrease as observed in previous studies, approaching control values during the next 30-60 min.

As noted in the figure, the % swelling in response to 40% hypotonic media is significantly less than the predicted peak value of approximately 167%, characteristic of a perfect osmometer. This degree of swelling is also less than seen previously with isotopic dilution methods in isolated cell suspensions (Ballatori and Boyer, Am. J. Physiol. 262:G451-G460, 1992). The reason for the decreased volume response is not clear although cells adherent to cover slips might have a restricted response compared to cells in suspension and may not behave as perfect spheres. Therefore this method may not reflect true volume response. Nevertheless this imaging technique allows independent data to be collected for single cells where measurements of changes in volume can be sequentially determined in real time. Supported by NSF ESI-9452682, ES-03828, DK-25636 and 34989.

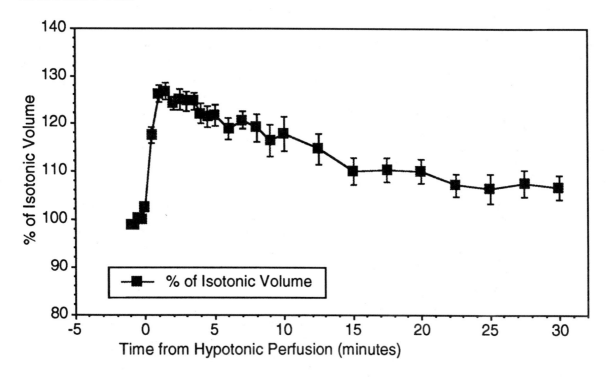

Fig.   Effect of Hypotonic Shock on change in volume (±SE) of single isolated skate hepatocytes (n=20) in response to perfusion with a 40% dilution of Elasmobranch Ringer's as assessed by an image analysis technique.

CHLORIDE AND TAURINE EFFLUXES OCCUR BY DIFFERENT PATHWAYS IN <u>RAJA ERINACEA</u> ERYTHROCYTES

Erin M. Davis-Amaral, Amaris Acosta, Elizabeth Wilson and Leon Goldstein
Physiology Department, Brown University, Providence, RI, 02912

We have been engaged in long-term study of volume-activated osmolyte (e.g. taurine) transport systems in skate RBC. The exact mechanism of this transport has yet to be determined and has sparked controversy. Our previous studies [Goldstein and Davis, Am. J. Physiol., 267:R426-R431, 1994] showed that volume-activated taurine transport occurs via a Na^+-independent, bi-directional transporter which has the properties of a size-limited channel. Evidence suggests that band 3 is involved in either formation or regulation of this channel. However, Kirk et al. [J. Biol. Chem., 267:23475-23478, 1992] believe that volume-activated Cl^- channels are involved in the transport of osmolytes across the fish RBC membrane. Their hypothesis is based upon studies with anion channel blockers. The aim of this study was to examine and to compare Cl^- efflux to taurine efflux in skate RBC under isotonic and hypotonic conditions.

Cl^-, K^+ and taurine effluxes were measured in isotonic and hypotonic media in which Cl^- and Na^+ were replaced by gluconate and mannitol; isotonic (940 mosmol/l) in mM: 510 mannitol, 2.7 Mg^{+2} D-gluconate, 5.0 Ca^{+2} D-gluconate, 15.0 Tris, 370 urea, pH 7.5; hypotonic (460 mosmol/l): mannitol was reduced to 185mM and urea was reduced to 250mM. All incubation media contained 0.1mM methazolamide to minimize HCO_3^- formation and Cl^-/HCO_3^- exchange. Final concentration of RBC in incubation media was 5%. ^3H-taurine was added to incubation media at final concentration of 0.1mM (2µCi/ml). Medium aliquots were taken and analyzed for Cl^- coulombmetrically, K^+ by flame photometry, and taurine efflux by liquid scintillation counting. Dry weight (DW) was obtained by heating a RBC aliquot overnight at 80°C. RBC pH was measured by freezing a RBC pellet in an EtOH/dry ice bath, the cells warmed in 15°C water bath and pH taken [T. McManus, personal communication].

Cl^- efflux showed no change in isotonic (mean± S.E.,(11) = 2.8±0.6 µmol/g•DW RBC•min) vs. hypotonic (3.5±0.9 -- P>0.05) stimulated RBC while taurine efflux rose from 0.045± 0.02 µmol/g•DW RBC•min (n=6) to 2.1±0.05 -- P<0.01. We sought to characterize the nature of Cl^- efflux. We found that there was no net K^+ efflux in either 940 or 460 media. Thus, Cl^- efflux was not accompanied by a conductive K^+ flux. To further clarify the nature of the Cl^- efflux, inhibitor studies were conducted. NPPB, quinine and arachidonic acid-- known inhibitors of volume-activated Cl^- channels-- had no significant effect on the Cl^- efflux, while DIDS-- a band 3 inhibitor-- completely inhibited Cl^- flux suggesting that Cl^- efflux was due to Cl^-/OH^- exchange. Therefore, the pH of the RBC was assayed after an hour incubation with and without DIDS. In the absence of DIDS, the pH went from (mean± S.E., n=6) 7.6±0.06 to 8.4±0.04 -- P<0.01, while in the presence of 0.1mM DIDS the pH rose only to 7.95±0.04 -- P<0.01. These results are consistent with Cl^- efflux via Cl^-/OH^- exchange. Bisognano et al. [J. Gen. Physiol., 102:99-123,1993] showed the presence in human RBC of Cl^-/OH^- exchange (or Cl^-/H^+ cotransporter). Similarly, their system was inhibited by DIDS. In conclusion, we showed that Cl^- flux is <u>not</u> volume-activated nor conductive. Taurine and Cl^- fluxes are apparently under different pathway influences: taurine diffuses via a channel while Cl^- is transported by exchangers (or cotransporters). Research supported by NSF: DCB 9102215 (L.G.) and ESI-9452682 (MDIBL).

ALTERATION OF ANKYRIN BINDING ACCOMPANIES VOLUME EXPANSION IN LITTLE SKATE (<u>RAJA ERINACEA</u>) ERYTHROCYTES

Mark W. Musch[1] and Leon Goldstein[2]

[1]Dept. of Medicine, Inflammatory Bowel Disease Center, Univ. of Chicago, Chicago, IL 60637
[2]Dept. of Physiology, Brown Univ., Providence, RI 02912

After volume expansion, the band 3 homolog in skate erythrocytes forms an altered allosteric state which forms tetramers. Under isoosmotic conditions, few tetramers are found and band 3 exists predominantly in a dimeric form. One functional difference between these forms of band 3 is their interaction with the cytoskeletal protein ankyrin. Ankyrin is a 210 kDa protein which forms one of the most important interactions of band 3 with the cytoskeletal structure which lies immediately beneath the erythrocyte membrane. Ankyrin has two functional binding domains, one for band 3 and another for spectrin (Bennett and Gilligan, Ann. Rev. Physiol. 9:27-66, 1993). The change in band 3 from the dimer to tetramer causes an increase in the affinity on ankyrin binding. Our goal was to determine whether such a change in ankyrin binding occurred in intact cells under conditions that stimulate tetramer formation physiologically.

Skate erythrocytes were volume expanded either by incubation in medium of one half osmolarity or by the inclusion of the permeant solute ethylene glycol. In addition, formation of band 3 tetramers was stimulated by including DNDS or pyridoxal-5-phosphate (PLP) under isosomotic conditions. For most experiments, since volume expansion is maximal near 5 minutes, cells were lysed at this point and ghosts rapidly prepared. A complete time course to determine changes over time was performed under hypotonic conditions to determine reversibility of any effects. After ghosts were isolated, peripheral membrane proteins were stripped by inclusion of potassium iodide in the lysis buffer, forming inside-out vesicles (KI-IOV).

Ankryin for the binding studies was isolated from human red cells using anion exchange chromatography (Bennett, Meth. Enzy,ol. 96:313-324, 1983). We were not able to purify skate ankryin to homogeneity using this procedure, perhaps due to the complexity of the skate erythrocyte. Skate ankyrin is recognized by antibodies to human ankyrin, and therefore must share some similarities. The human ankyrin was iodinated (^{125}I using Bolton-Hunter kit; Pierce) and used in binding studies. Experiments were conducted as soon as possible after isolation of KI-IOV and were allowed to proceed for only 60 min since prolonged exposure spontaneously leads to changes in membrane proteins in the KI-IOV. Care was taken to keep the pH at 7.5 until the binding studies since higher pH values also promote spontaneous "aggregation" of membrane proteins. Binding characteristics were calculated from Scatchard analysis using the model presented by Thevinin and Low (J. Biol. Chem. 265: 16166-16172, 1990)

Under volume expanded conditions, the total number of ankyrin binding sites remained constant ,but demonstrated a shift in affinity. Under isoosmotic conditions, nearly all of the binding was in a population with an affinity of approximately 140nM. Less than 5% demonstrated an affinity of 35nM. Incubation in either hypotonic or ethylene glycol media stimulated an increase in the high affinity population. In three separate experiments, the shift to the high affinity class was 55% for hypotonicity and 40% for ethylene glycol. Incubation with DNDS or PLP also stimulated a shift to the high affinity sites of about 35% and did not increase the number of binding sites. To determine if this was due to interaction with skate KI-IOV band 3, the cytoplasmic domain of human band 3 was purified. When included in the incubation medium in excess of the predicted number of binding sites, 80% of the high affinity sites could be competed away. The results suggest that volume expansion leads to an alteration of the interaction of band 3 with the cytoskeleton through one of its most important links, ankyrin. This interaction may be pivotal in controlling the activity of this transport protein under volume expanded conditions.

Supported by NSF grant DCB 9102215 and NIH grants DK38510, 42086, and 47722.

PYRIDOXAL BINDING SITES IN VOLUME EXPANDED LITTLE SKATE RAJA ERINACEA ERYTHROCYTES

Mark W. Musch[1], Erin M. Davis-Amaral[2], and Leon Goldstein[2]
[1]Dept. of Medicine, Inflammatory Bowel Disease Center, Univ. of Chicago, Chicago, IL 60637
[2]Dept. of Physiology, Brown Univ., Providence, RI 02912

After volume expansion, efflux of the ß-amino acids increases dramatically in skate erythrocytes, as well as many other cells. Taurine transport occurs via a number of pathways and to investigate potential membrane transporters, an inhibitor which may be covalently linked to membrane proteins was used. Pyridoxal-5-phosphate (PLP) is a potent inhibitor of volume-activated taurine efflux. PLP reacts with amino groups of proteins to form a Schiff base. In the presence of sodium borohydride, this association can be made covalent. Therefore we used [^{3}H]-borohydride to label membrane proteins which are reactive with PLP and which may be the transporter for taurine or an associated regulatory protein (Cabantchik et al., J. Biol. Chem. 250: 5130-5136, 1975). To confirm specificity, concentration-response relationships were determined as well as the ability of other inhibitors of taurine efflux to alter the pattern of PLP binding.

Erythrocytes were incubated in isotonic or hypotonic elasmobranch incubation medium alone or in the presence of one of the following taurine transport inhibitors: arachidonic acid (50μM) and its non-active analog arachidic acid (50μM), DNDS (0.5mM), DIDS (0.5mM), quinine (1mM), pyridoxal (0.2 and 2mM), and NPPB (0.5mM). The kinetics of PLP binding were determined in the absence of other inhibitors. Binding occurred rapidly and increased over time. Binding at 10 min was approximately 50% compared with that at 60 min. This time point was used in subsequent inhibitor studies. Using 1mM sodium borohydride (2200cpm/nmole), no labelling occurred. Binding could be observed with concentrations of PLP as low as 0.02mM; more binding was observed at 0.2mM PLP and at 2mM PLP, labelled a large number of membrane proteins were labelled. The three primary proteins labelled at 0.2mM were at 100, 80, and 65kDa. The peak at 100kDa was likely the skate homolog of band 3, but the identity of the others is unknown. DNDS, DIDS, and NPPB inhibited PLP binding at the 80kDa protein (as well as at band 3). However, little inhibition of PLP binding was observed in the 65kDa protein. Additionally, the potent transport inhibitors arachidonic acid and quinine did not inhibit the binding at any peak. Pyridoxal labelled nearly the same proteins at PLP and followed the same inhibitor profile as PLP. Cleveland digest analysis of the PLP-labelled proteins with trypsin, chymotrypsin, and papain demonstrate that the 80kDa protein is unique. Characterization of this protein may identify a taurine transporter which is regulated during volume expansion.

Supported by NSF grant DCB 9102215 and NIH grants DK38510, 42086, and 47722.

VANADATE INHIBITION OF ORGANIC ANION SECRETION IN KILLIFISH, <u>FUNDULUS HETEROCLITUS</u>, RENAL PROXIMAL TUBULES

Luana Atherly[1] and David S. Miller[2]
[1]Lincoln University, Lincoln University, PA 19352
[2]Lab. of Cellular and Molecular Pharmacology
NIH-NIEHS, Research Triangle Park, NC 27709

The organic anion transport system in vertebrate renal proximal tubule mediates the transport from blood to urine of a large number of potentially toxic metabolic wastes, xenobiotics and xenobiotic metabolites. Membrane vesicle studies have suggested that secretion of organic anions is driven by indirect coupling to the Na-gradient at the basolateral membrane and PD-dependent facilitated diffusion at the luminal membrane (Pritchard and Miller, Physiol. Rev. 73:765, 1993). Indirect coupling of organic anion influx to Na has been demonstrated in many intact tissue preparations, but little is known about how organic anions move from cell to lumen in intact renal tubules. Last summer, work from this laboratory demonstrated that the transport of several fluorescent organic anions across the lumenal membrane of killifish proximal tubule was not reduced when cells were depolarized in high potassium medium (Miller et al., Am. J. Physiol., submitted). We concluded that a mechanism other than PD-driven facilitated diffusion was responsible for transport across the lumenal membrane and suggested that, in analogy to liver, an organic anion transporting ATPase might be involved. Here we report on initial experiments with vanadate, a potent inhibitor of p-type ATPases.

Renal tubular masses were isolated in a marine teleost saline (MTS; containing, in mM:140 NaCl, 2.5 KCl, 1.5 CaCl$_2$, 1.0 MgCl$_2$ and 20 tris(hydroxymethyl)-amino methane, at pH 8.25). Under a dissecting microscope masses were teased with forceps to remove adherent hematopoietic tissue. Individual killifish proximal tubules were dissected free of the masses and transferred to a foil-covered Teflon chamber (Bionique) containing 1 ml of MTS with 1 μM fluorescein (FL), a fluorescent substrate for the organic anion transport system. The chamber floor was a 4x4 cm glass cover slip to which the tubules adhered lightly and through which the tissue could be viewed by means of an inverted microscope equipped with epi-fluorescence optics and a video camera connected to a Macintosh computer (Miller et al., Am. J. Physiol. 264:R882, 1994).

Previous imaging studies with killifish tubules have shown rapid uptake and secretion of FL into the tubular lumen; steady state is attained within 15 min (Miller et al., Am. J. Physiol. 264:R882, 1994). At steady state, lumenal fluorescence averages 2-3 times cellular fluorescence. Figure 1 shows that addition of vanadate to the incubation medium caused a concentration dependent increase in both cellular and lumenal fluorescence. However, with increasing vanadate concentration, cellular fluorescence increased more rapidly than lumenal fluorescence, so the lumen to cell fluorescence ratio for FL fell from a control value of 3.0 ± 0.4 to 1.1 ± 0.1 with 10 μM vanadate. Additional experiments revealed significant decreases in this ratio with vanadate concentrations as low as 0.5 μM, but with longer exposures (not shown). Time course studies showed that 5 μM vanadate increased cellular fluorescence within 10 min, but that the increase in lumenal fluorescence required 20-30 min of exposure.

Vanadate has been shown to inhibit renal Na,K-ATPase (Nechay, Ann. Rev. Pharmacol. Toxicol. 24:501, 1984). To determine whether the observed pattern of vanadate effects on FL transport could be explained by Na,K-ATPase inhibition, we examined the effects of ouabain, a specific inhibitor of Na,K-ATPase, on FL transport by killifish tubules. In contrast to vanadate, 1-100 μM ouabain caused a concentration dependent decrease in lumenal and cellular fluorescence (not shown). With 100 μM ouabain, a concentration that completely inhibits teleost renal Na,K-ATPase (Miller, J. Pharmacol. Exp. Therap. 219:428, 1981), FL uptake into the cells and secretion into the lumen were abolished. This is the expected result, since Na,K-ATPase inhibition should collapse the Na gradient that drives organic anion uptake by the cells, which should in turn inhibit transport into the lumen.

The results of the present study show that low concentrations of vanadate have multiple effects on organic anion transport by killifish renal proximal tubules as measured by the distribution of a fluorescent substrate, FL; these effects are not related to Na,K-ATPase inhibition. Vanadate significantly increased both cellular and lumenal fluorescence. It is important to note that with the present nonconfocal optics lumenal fluorescence is expected to increase when cellular fluorescence increases, since at least a portion of the fluorescence measured over the lumen is contributed by the cells above and below. The lumen to cell fluorescence ratio allows us to semiquantitatively correct lumenal values for changes in cellular fluorescence. It is significant that this ratio decreased even though fluorescence in both compartments increased. The fluorescence ratio data suggest that vanadate inhibited transport of FL from cell to lumen. The observed increase in cellular FL could be secondary to inhibition of lumenal exit or it might indicate a separate effect on the basolateral transport mechanism, e.g., inhibited efflux. Supported in part by a fellowship from Burroughs-Welcome and NSF REU 9322221.

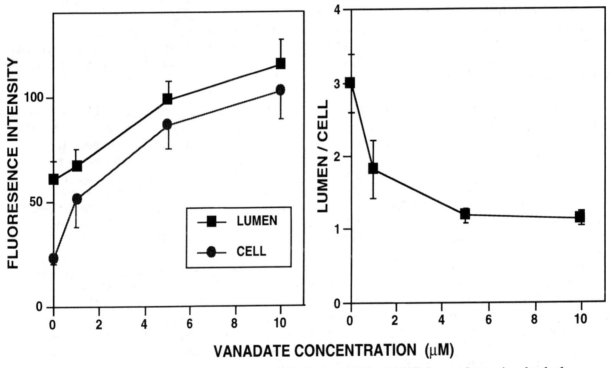

Figure 1. Effects of vanadate on fluorescein (FL) transport by killifish renal proximal tubules. Tissue was incubated in medium with 1 μM FL and the indicated concentration of vanadate for 30 min. Data given as mean ± SE for 24-32 tubules. All concentrations of vanadate significantly increased cellular fluorescence and significantly decreased the lumen to cell fluorescence ratio ($P < 0.05$ for 1 μM vanadate and $P < 0.01$ for 5-10 μM); 5 and 10 μM vanadate significantly increased lumenal fluorescence, $P < 0.01$.

EXTRACELLULAR ATP AND Hg^{2+} MOBILIZE Ca^{2+} FROM DISTINCT INTRACELLULAR POOLS IN HEPATOCYTES ISOLATED FROM THE LITTLE SKATE RAJA ERINACEA

Michael H. Nathanson and Kavita Mariwalla

Liver Study Unit, Yale University School of Medicine, New Haven, CT 06520

We have previously shown that skate hepatocytes contain ATP receptors, and that stimulation of these receptors increases cytosolic Ca^{2+} (Ca$_i^{2+}$)(M.H. Nathanson and K. Mariwalla, Amer. J. Physiol. (in press), 1996). This Ca$_i^{2+}$ increase consists of two components: an early phase due to release of internal Ca^{2+} stores, which reaches its peak within seconds, and a late, prolonged phase due to influx of extracellular Ca^{2+}, which persists for minutes. We also have shown that Hg^{2+} increases Ca$_i^{2+}$, although this Ca$_i^{2+}$ increase is dose-dependent and predominantly due to release of Ca^{2+} from internal stores (M.H. Nathanson et al, Cell Calcium 18:429-439, 1995). The purpose of this study was to further characterize the receptors and internal stores responsible for ATP-induced Ca$_i^{2+}$ signals, and to determine the relationship between these Ca^{2+} stores and those mobilized by Hg^{2+}. Hepatocytes were isolated by collagenase perfusion (D.J. Smith et al, Amer. J. Physiol. 252:G479-G484, 1987), then either loaded with the Ca^{2+}-sensitive dye indo-1 (10 μM) and examined by ratio spectrofluorometry (G. Grynkiewicz et al, J. Biol. Chem. 260:3440-3450, 1985) using a Perkin-Elmer LS-5B spectrometer, or loaded with the mitochondrial dye rhodamine-123 (5 μg/ml) and examined by confocal fluorescence microscopy.

Since there is variability in the response of skate hepatocytes to ATP, we first examined whether this variability was due in part to release of endogenous ATP, leading to desensitization of purinoceptors. We compared the response of control hepatocytes to hepatocytes pre-incubated with the ATP/ADPase apyrase. In control hepatocytes, the K$_m$ for ATP was 1.2±1.0 μM (mean±SEM), and in hepatocytes pre-incubated with apyrase the K$_m$ was not significantly different (1.6±1.1 μM); each K$_m$ was estimated by nonlinear regression from dose-response curves in which each data point was measured in duplicate or triplicate. Thus, apyrase did not alter the dose-response relationship for ATP, suggesting that desensitization of receptors due to release of endogenous nucleotides does not contribute to the variable response we observed. We also performed cross-desensitization studies to further characterize skate hepatocyte ATP receptors. Cells were stimulated sequentially with UTP (100 μM) followed by 2MeSATP (100 μM), or else by 2MeSATP followed by UTP. Stimulation with UTP initially increased Ca$_i^{2+}$ by 415±85 nM, and subsequent exposure to 2MeSATP further increased Ca$_i^{2+}$ by only 83±90 nM (n=3; p<0.02 by paired t test). In contrast, initial stimulation with 2MeSATP increased Ca$_i^{2+}$ by 329±18 nM, while subsequent exposure to UTP further increased Ca$_i^{2+}$ by 537±103 nM (n=3; p>0.05 by paired t test). Thus, UTP desensitized hepatocytes to 2MeSATP, while 2MeSATP did not similarly affect the response to UTP.

To investigate the relationship between internal Ca^{2+} stores mobilized by ATP and Hg^{2+}, cells were placed in Ca^{2+}-free medium, then sequentially stimulated with ATP (100 μM) to mobilize agonist-sensitive Ca^{2+} stores, the Ca^{2+}-ATPase inhibitor thapsigargin (5 μM) to further deplete these stores, and Hg^{2+} (50 μM). Each of these additions increased Ca$_i^{2+}$ (p<0.0001, p<0.02, and p<0.00005, respectively). These Ca$_i^{2+}$ increases also occurred if thapsigargin was added before ATP (n=4). This suggests that ATP and Hg^{2+} mobilize Ca^{2+} from distinct internal stores. Mitochondria provide a Ca^{2+} pool that is not mobilized by agonists such as ATP. Since mitochondrial sequestration of Ca^{2+} is dependent upon the maintainence of a highly negative mitochondrial membrane potential (T.E. Gunter et al, Amer. J. Physiol. 267:C313-C339, 1994), we examined the effect of Hg^{2+} on this potential. Isolated skate hepatocytes were loaded with the cationic fluorescent dye rhodamine-123, which accumulates in mitochondria in direct proportion to their membrane potential (A.L. Nieminen et al, J. Biol. Chem. 265:2399-2408, 1990). Mitochondrial fluorescence over time was measured by time-lapse confocal microscopy in the absence or presence of Hg^{2+} (Figure 1). Hg^{2+} accelerated the loss of fluorescence from mitochondria, suggesting that Hg^{2+} decreases the potential gradient, permitting mitochondrial Ca^{2+} to leak into the cytosol.

In summary, these findings suggest: (1) skate hepatocytes express ATP receptors which exhibit broad specificity, (2) stimulation of ATP receptors and application of Hg^{2+} mobilize Ca^{2+} from distinct internal stores, and (3) Hg^{2+} dissipates the potential gradient of mitochondria, suggesting that mitochondria are the source of Hg^{2+}-induced Ca_i^{2+} signals in this cell type. Additional work will be needed to demonstrate directly that depolarization induces Ca^{2+} release from mitochondria, and to define the role of mitochondrial depolarization in Hg^{2+}-induced toxicity in skate hepatocytes.

This work was supported by a Young Investigator Award (to MHN) from the Center for Membrane Toxicity Studies (P30 ES03828), a Student Research Award (to KM) from the American Gastroenterological Association, a Liver Scholar Award from the American Liver Foundation (to MHN), a Grant-in-Aid from the American Heart Association (to MHN) and the Hepatocyte Isolation and Morphology Core Facilities of the Yale Liver Center (P30 DK34989).

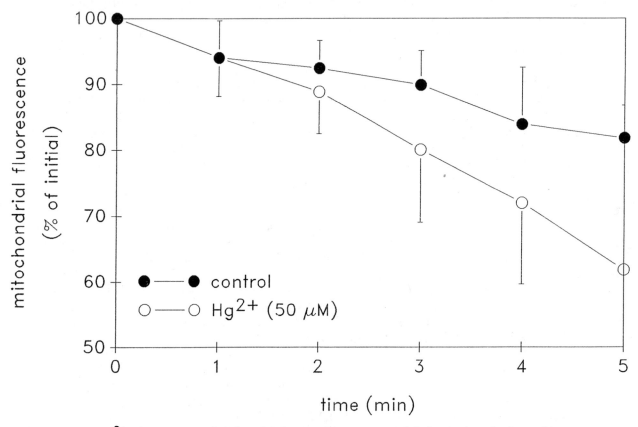

Figure 1. Hg^{2+} decreases mitochondrial membrane potential in isolated skate hepatocytes. Hepatocytes were loaded with the fluorescent dye rhodamine-123, which accumulates in mitochondria in proportion to mitochondrial membrane potential. Mitochondrial fluorescence decreases slowly under control conditions (solid circles), and rapidly in the presence of Hg^{2+} (open circles).

ATP CONTENT AND RELEASE IN CULTURED RECTAL GLAND CELLS FROM THE SHARK SQUALUS ACANTHIAS

Horacio F. Cantiello[1], Carlin F. Jones[2], Adriana G. Prat[1], George R. Jackson Jr.[1]

[1]Renal Unit, Massachusetts General Hospital East, Charlestown, MA 02129 and Department of Medicine, Harvard Medical School, Boston, MA 02115

[2]Lake Erie College, Painesville, OH 44077

Recent studies from our laboratory have shown that the cystic fibrosis transmembrane conductance regulator (CFTR) is capable of conducting ATP as the charge carrier (Reisin et al., J. Biol. Chem., 269: 20584-20591, 1994). The physiological role of the ATP-conductive pathways may be associated with the autocoid delivery of ATP to the extracellular milieu, and the consequent stimulation of the outwardly rectifying chloride channel (ORCC, Schwiebert, et al. Cell 81: 1063-73, 1995). The CFTR-related protein, P-glycoprotein, responsible for the multidrug resistance phenotype in tumors, is also an ATP channel (Abraham et al., Proc. Natl. Acad. Sci. USA, 90: 312-316, 1993).

Little information is available on the ATP-permeable pathways of the shark. We recently determined that cultured shark rectal gland (SRG) cells express a cAMP-activated electrodiffusional pathway that is permeable to both ATP and Cl (Cantiello et al., Bull. MDIBL 33: 47-48, 1994). Although the ATP-conductive pathway of SRG cells was activated by cAMP, a behavior expected for a CFTR-like molecule, this pathway showed rectifying properties in symmetrical ATP, and a pharmacological profile (resistance to DPC, inhibition by nifedipine) more similar to that of the murine isoform of the P-glycoprotein. In this report we determined the intracellular ATP content and the characteristics of ATP release in SRG cells using the luciferin-luciferase assay.

Primary cultures of shark rectal gland cells were obtained from adult male Squalus acanthias, as previously described (Valentich, Bull. MDIBL 26: 91-94, 1986). Cells were seeded at a high density for primary cultures to spread onto coverslips. Extracellular and total cellular ATP were measured with a modified luciferin-luciferase assay as previously described (Abraham et al., op.cit.). Briefly, at the time of the experiment, the coverslips were placed in plastic cuvettes and held vertically in the cuvette with a microclip (Roboz Surgical Instruments, Inc., Rockville, MD). The assay solution contained 0.1 ml of the luciferin-luciferase assay mix (Sigma Chem. Co., St Louis, MO) and 0.5 ml of a Ca^{2+}-free solution containing 280 mM NaCl and 10 mM Hepes, pH 7.4. Whenever indicated, NaCl was replaced with Na-gluconate. Photon release was continuously measured in a luminometer (MonoLight 2010, Analytical Luminescence Lab., Ann Arbor, MI). The ATP release was determined by the photon release of the luciferin-luciferase assay for ~2 min (to reach a steady-state plateau) before membrane permeabilization to assess the total intracellular ATP. Permeabilization was accomplished by addition of alamethicin (10 µM) and sonication (30 sec, Ultrasonic sonicator, FS-14, Fisher Scientific). Photon release was again followed for another 2 min. To determine the amount of ATP released from cells, known concentrations of

ATP in solution were also measured to construct a calibration curve. The average ratio of extracellular to total ATP was obtained after the background signal level was subtracted and the ATP released was compared to total ATP content obtained in each case. Whenever indicated, cells were incubated overnight in the presence of cholera toxin (CTX, 6 µg/ml, Sigma).

Intracellular ATP content was measured after complete permeabilization with alamethicin as indicated above (Fig. 1). Intracellular ATP was 200% higher in Cl-free (gluconate) solution (35.2 ± 8.4, pmoles ATP/coverslip, n=13, vs. 107 ± 29.2, pmoles ATP/coverslip, n=8, p<0.05, for Cl and gluconate solutions, respectively). Thus, intracellular ATP was dependent on the presence of Cl in the extracellular milieu. Under these (physiological) conditions, however, no significant cAMP-dependent change in the ATP content was observed, at least according to the lack of effect of cholera toxin (Fig. 1), since the toxin-treated cells also had proportionally higher ATP in the Cl-free solution (289%, p<0.005).

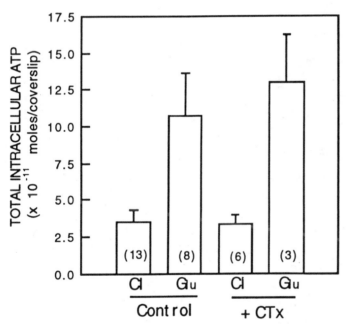

Fig. 1. Effect of cholera toxin (6 µg/ml) on ATP content of cultured SRG cells. ATP content was measured in 280 mM NaCl (Cl) or a Cl-free solution (Glu) where NaCl was replaced with Na-gluconate. Values were obtained after complete permeabilization of cells with alamethicin. Values are the mean ± SEM for experiments indicated in the bars.

Whenever the rate of ATP release (Prat et al., Am. J. Physiol., 270: In press, 1996), was determined as the extracellular to total ATP ratio, it was observed that no significant differences were found (Fig. 2) between Cl-containing and Cl-free solutions for control and CTX-treated cells. However, the ATP ratio was 63.5% lower in CTX-treated cells in the Cl-free solution. Although this difference did not reach statistical significance, the data are suggestive of a possible cAMP- (CTX)-dependent inhibition of ATP release in the absence of extracellular Cl (Fig. 2). Because the rate of ATP release did not change in the presence of external gluconate (Fig. 1), the data suggested that the higher ATP contents in the Cl-free solution may be only partially explained by changes in the rate of ATP release. This is further supported by the data of Fig. 3, where the rate of ATP release and intracellular ATP were inversely related, albeit independent of treatment with CTX. The data indicate that intracellular ATP is inversely proportional to the rate of ATP release. The correlation (Natural log-Linear) is consistent, however, with a constitutive mechanism for the release of cellular ATP, and which is not activated by cAMP.

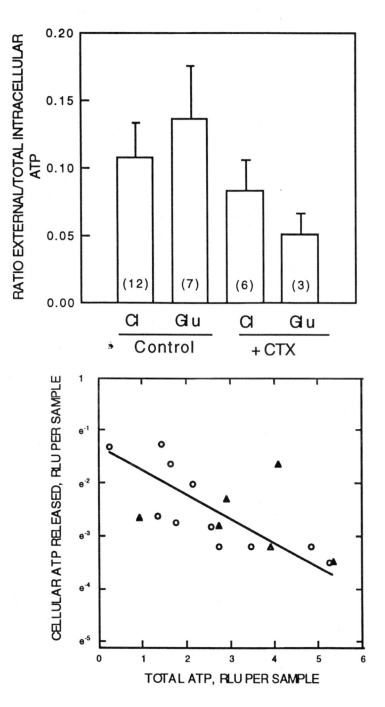

Fig. 2. Effect of cholera toxin (6 µg/ml) on ATP released by of cultured SRG cells. Data are expressed as the ratio between external vs. total ATP. ATP content was measured in 280 mM NaCl (Cl) or a chloride-free solution (Glu) where NaCl was replaced with Na-gluconate. Values of total intracellular ATP were obtained after complete permeabilization of cells with alamethicin. Values are the mean ± SEM for experiments indicated in the bars.

Fig. 3. Correlation between cellular ATP released and total ATP. Data are expressed in arbitrary light units (RLU). Values were obtained for control (circles) and cholera toxin-treated SRG cells (triangles) in chloride-free solution. Ordinate axis is a natural log of RLU, thus indicating the exponential correlation between released ATP and total nucleotide content.

The data in the present study indicate that the maintenance of intracellular ATP content of cultured SRG cells is largely dependent on replacing the extracellular anion, which is most consistent with the presence of a constitutive (not cAMP-stimulated) ATP pathway which elicits the spontaneous release of cellular ATP. This is in contrast to our previous patch-clamp observations indicating that cAMP-stimulation activates a pathway that is permeable to both Cl and ATP. However, the cells were largely insensitive to CTX. Although this may be due to a possible resistance of shark cells to cholera toxin, SRG cells were also insensitive to cAMP analogs, isobutyl-methyl-xanthine, and forskolin (data not shown), thus indicating that the batches of cells used in this study failed to express functional "CFTR-like" ATP pathways. However, a basal ATP pathway was also observed in our previous studies, thus indicating that putative ATP transporters other than CFTR may also be present. In this regard, the data on the decrease of the rate of release of ATP by CTX on the Cl-free-bathed cells may be suggestive of a transport mechanism that is likely to be down-regulated by cAMP-stimulation. Although this hypothesis will require further studies to be confirmed, recent evidence indicates that stimulation of the cAMP pathway down-regulates the expression of P-glycoprotein (Scala et al., J. Clin. Invest., 96: 1026-34, 1995).

An interesting aspect of the data in this report relates to the fact that in extracellular Cl, SRG cells had much lower intracellular ATP than in Cl-free solution. This may be explained by the competition between cellular ATP and Cl for transport through the same transport mechanism. CFTR, for example, is able to move either one of the anions. However, previous studies on P-glycoprotein indicate that this channel only moves ATP and is impermeable to Cl. The higher ATP content in external gluconate, therefore, may be associated with a higher electrochemical Cl gradient which then disallows ATP from coming out, and/or the possibility that in the absence of external Cl, a major contributor to ATP movement, the resting potential of the cell may have changed. Considering the depolarizing effect of external Cl replacement, the possibility exists for a change in the electrochemical driving force for cellular ATP movement. This is supported by recent studies where it was calculated that the resting potential of the cell is perhaps the most relevant acute regulator of ATP release (Prat et al., Am. J. Physiol., 270: In press, 1996). This is also consistent with the fact that the rate of ATP release was not modified, thus suggesting that it is not the transport mechanism itself, but the driving force associated with the ATP movement which is modified. The data argue in favor of a potential electrodiffusional ATP pathway, at least in the context of its sensitivity to cell depolarization. This constitutive pathway for ATP release may play a relevant role in the regulation of the epithelial cell response associated with adenosine nucleotide derivatives in SRG cells.

The authors gratefully acknowledge Dr. John Forrest, Jr. for providing the cultured SRG cells. Studies were funded by grants from the Cystic Fibrosis Foundation and the Center for Membrane Toxicity Studies. CFJ was supported by the American Heart Association, Maine Affiliate.

EVIDENCE FOR AN ATP-CONDUCTIVE PATHWAY IN RETINAL NEURONS OF THE SHARK SQUALUS ACANTHIAS

Horacio F. Cantiello[1], George R. Jackson Jr.[1], Greg Maguire[2], and Harvey Karten[3]

[1]Renal Unit, Massachusetts General Hospital East, Charlestown, MA 02129 and Department of Medicine, Harvard Medical School, Boston, MA 02115

[2]Sensory Sciences Center, University of Texas, Houston TX, 77030

[3]Department of Neurosciences, University of California at San Diego, La Jolla, CA 92093

ATP acts as a neurotransmitter in both the central (Edwards, et al., Nature 359: 144-7, 1992) and peripheral (Evans, et al., Nature 357: 503-5, 1992) nervous systems. Recent studies have also demonstrated that external ATP is associated with P_2 purinergic receptor activation in the retina (Kirischuk, et al., J. Physiol., 483: 41-57, 1995). It is possible, therefore, to suggest that brain tissues, and in particular the retina, may have specific transport mechanisms to elicit the release of ATP, which may then act as an autocoid regulator of neuronal function.

Recent studies have shown that the cystic fibrosis transmembrane conductance regulator (CFTR), an anion channel which is capable of conducting ATP as the charge carrier (Reisin et al., J. Biol. Chem., 269: 20584-20591, 1994), is expressed in mammalian brain tissues (Mulberg, et al., J. Neurochem., 64: 1662-8, 1995). The role of CFTR in the central nervous system, however, is as yet unknown. We have recently determined that cultured shark rectal gland (SRG) cells express a cAMP-activated electrodiffusional pathway that is permeable to both ATP and Cl⁻ (Cantiello et al, Bull. MDIBL 33: 47-48, 1994). Although the ATP-conductive pathway of SRG cells was activated by cAMP, a behavior expected for a CFTR-like molecule, this pathway showed rectifying properties in symmetrical ATP, and resistance to the CFTR-channel blocker diphenylamine carboxylic acid (DPC), and the Cl⁻ channel blockers DIDS, and anthracene-9- carboxylic acid (9AC), yet it was readily blocked by nifedipine (500 µM). In this report we used patch-clamp techniques to determine the presence of an ATP-conductive pathway in isolated shark retinal neurons.

Whole eyes were dissected from double pithed adult sharks. The cornea was dissected and the retina was removed with a spatula and transferred to a petri dish. The isolated retinas were minced and incubated for 30 min at room temperature in a papain solution (1mg/ml) in 280 mM NaCl buffer. 200 µl of the suspension was then placed into the patch clamp chamber and isolated neurons were allowed to attach for up to 30 min. The chamber was then perfused with saline to remove unattached neurons.

Whole-cell, cell-attached, and excised inside-out patch clamp techniques were applied to bipolar cells placed in a bathing solution containing 280 mM NaCl. The patch pipette was filled with 200 mM MgATP, pH 7.4. Under whole-cell conditions, asymmetrical ATP/Cl⁻ currents increased 1,440% with the addition of 1 mM cyclophenylthio-cAMP (cpt-cAMP) (22.8 ± 0.35 nS/cell, n=3 vs. 1.48 ± 0.10, n=3, p<0.001, Fig. 1). Whole cell currents were highly linear (r=0.9991), thus indicating that the cAMP-stimulated pathway was permeable to both Cl⁻ (positive currents) and ATP (negative currents).

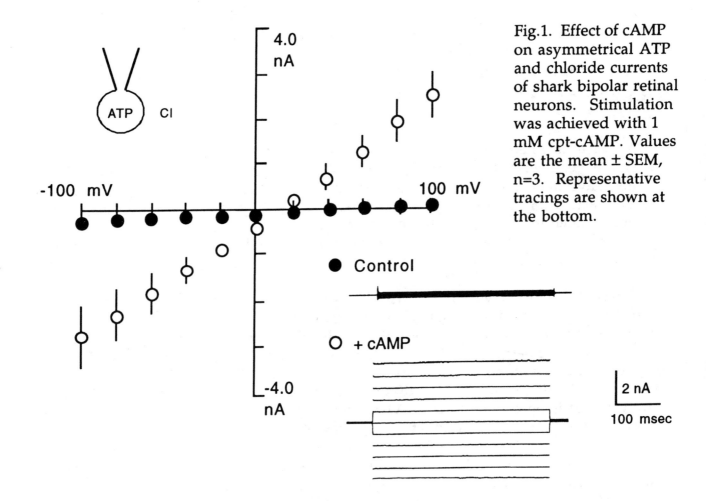

Fig.1. Effect of cAMP on asymmetrical ATP and chloride currents of shark bipolar retinal neurons. Stimulation was achieved with 1 mM cpt-cAMP. Values are the mean ± SEM, n=3. Representative tracings are shown at the bottom.

Under cell-attached and excised, inside out conditions, spontaneous ATP channels were observed in all cases (9 and 5 experiments, respectively, Fig. 2). Channel activity was insensitive to the anion channel blockers diphenylamine carboxilic acid (DPC), glibenclamide, and also resistant to nifedipine.

The data indicate that shark retinal neurons express an ATP-permeable channel sharing some functional similarities with a similar pathway in the rectal gland, namely ATP permeability and activation by cAMP. Immunocytochemistry studies with anti-CFTR antibodies (#13-1 or #24-1, Genzyme Corp. Framingham MA, data not shown) showed a strong labeling of amacrine cells and the inner plexiform layer, a region of synaptic junctions between bipolar cells and the ganglion cell layer. This finding suggests the presence of a protein with structural homology to a known ATP transporter, CFTR (Reisin, et al., op. cit.). However, the retinal ATP pathway was insensitive to anion channel blockers known to inhibit mammalian CFTR. Interestingly, the horizontal cell layer, which facilitates synaptic junctions between

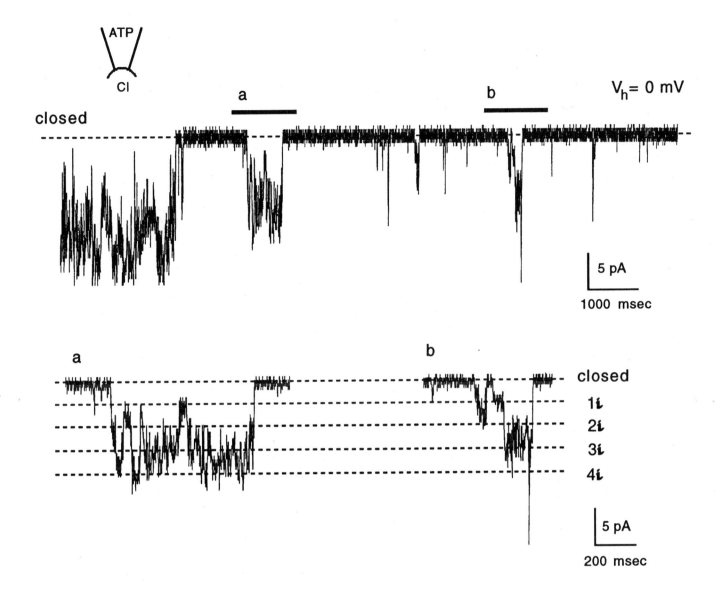

Fig. 2. Single channel ATP currents of shark bipolar neurons. ATP currents were spontaneously activated after patch excision. Pipette solution was 200 mM MgATP, and bathing solution contained 280 mM NaCl. Expanded tracings on bottom represent portions of top tracing underlined by the solid lines indicated (a & b). Tracings are representative of 5 experiments.

individual photoreceptors, stained strongly with an anti-P-glycoprotein antibody (Ab-1, Oncogene Science, Uniondale, NJ), suggesting the presence of another ATP transporter (Abraham, et al., op. cit.) structurally distinct from that in the lower retina. The focus of this study, however, was on bipolar cells, which form junctions between the inner and outer plexiform layers.

The ATP pathway in these cells is most consistent with the description of an ATP transport mechanism similar to that of mammalian CFTR. Clearly this needs to be further explored since it is possible that more than one ATP pathway may exist in the retina. The data in this report however, indicate the presence of an as yet unknown ATP pathway whose function and physiological relevance may be associated with the delivery of cellular ATP in the retina.

The studies were funded by grants from the Cystic Fibrosis Foundation and the Center for Membrane Toxicity Studies.

VOLUME-ACTIVATED TAURINE EFFLUX
FROM SQUALUS ACANTHIAS RECTAL GLAND CELLS

J. K. Haynes[1], Damien Sanderlin[1], Osak Omulepu[1], R. Lehrich[2], and J. N. Forrest, Jr.[2]
[1]Department of Biology, Morehouse College
Atlanta, Georgia 30314
[2]Department of Internal Medicine, Yale University
School of Medicine, New Haven, Connecticut 06510

Taurine efflux is one of the principal mechanisms responsible for regulatory volume decrease (RVD) in elasmobranch tissues (Boyd, T. et al., J. Exp. Zool. 199:435-442, 1977). Skate erythrocytes and liver cells as well as dogfish rectal gland cells have been the most frequently studied cell systems. While the mechanism of taurine efflux is relatively well-characterized in skate erythrocytes (Haynes, J. et al., Am. J. Physiol. 265:R173-R179, 1993) and liver cells (Ballatori, N. et al., Am. J. Physiol. 267:G285-G291, 1994), little is known about the transport mechanism in rectal gland cells (Ziyadeh, F. N. et al., Am. J. Physiol. 262:F468-F479, 1992). Regulation of taurine transport is poorly understood in all of the above systems.

The purpose of this investigation was to further clarify the mechanism of taurine transport during RVD in rectal gland cells and to determine if tyrosine phosphorylation plays a role in regulation of taurine transport. A primary monolayer cell culture system for shark rectal gland tubular epithelium provided an ideal system for further characterization of the taurine transport mechanism (Valentich, J. et al., Am. J. Physiol. 260:C813-C823, 1991).

One (1) to fourteen (14) day old rectal gland cultures from Squalus acanthias loaded with taurine by incubation with $1\mu c/ml$ of ^{3}H-taurine (1mc/ml) in 1 ml of shark rectal gland tissue culture (SRGC) medium for 2 hours in 12 well plates at 15°C in a CO_2 incubator (5% CO_2). The cells in each well were then washed 3X with SRGC. The efflux of taurine was measured after a 15 min incubation at 15°C in isotonic (1000 mOsM) and hypotonic (600 mOsM) SRGC medium, in which Na^+ was replaced by Li^+. The amount of isotope in the cytosol of cells was determined at 0 time and 15 minutes after lysing cells in buffer containing 1% NP-40, $150\mu M$ NaCl, $20\mu M$ Tris base (pH 8), $5\mu M$ EDTA, $10\mu M$ NaF_2, $10\mu M$ NaPPi, $10\mu g/ml$ aprotinin, $10\mu g/ml$ leupeptin, $10\mu M$ Iodocetamide, $1\mu M$ PMSF, and $1\mu M$ NaVanadate. The lysate was subjected to centrifugation at 15,000 rpm in a microfuge for 30 minutes to generate a post-nuclear supernatant fraction. 0.2 ml of supernate (in duplicate) was added to 4.0 ml Optiflour for scintillation counting. Proteins in the supernatant fraction containing phosphorylated tyrosine residues were detected in the following way: The supernate was diluted 1:1 with sample buffer, containing $62.5\mu M$ Tris HCl (pH 6.8), 2% SDS, 10% glycerol, 0.01% bromphenol blue, and β-mercaptoethanol. The sample was boiled for 7 minutes and subsequently subjected to SDS-PAGE (7.5% acrylamide). After electrophoresis, proteins were electrotransferred to Immobilon-P transfer membranes and subsequently probed with a monoclonal anti-phosphotyrosine antibody (4G10).

Rectal gland cells lost greater than 50% of intracellular taurine within 15 minutes when suspended in hypotonic medium, and the rate of efflux was substantially reduced when cells were shifted from hypotonic to isotonic medium after 5 minutes of incubation in the former medium (Fig. 1), suggesting that increased taurine efflux versus that observed in isotonic medium was an osmoregulatory response. While we have not yet determined whether taurine is metabolized during the loading period, previously published work indicates little or no metabolism of taurine during loading of a variety of elasmobranh tissues (King, P. et al. Mol. Physiol. 4:53-66, 1983). Thus, we interpret the loss of radioisotope from the cell as a measure of taurine efflux. In order to further characterize the transport mechanism, taurine efflux was measured in the presence and absence of the anion transport inhibitors: pyridoxal phosphate ($2\mu M$), DIDS ($0.5\mu M$) and NPPB ($0.1\mu M$).

None of these inhibitors had a significant effect on taurine efflux, even when cells were preincubated for up to 15 minutes in the inhibitor, suggesting that the transport pathway does not involve an anion channel or Band 3 (Figs 2 and 3; data on NPPB not shown).

Five (5) prominent phosphorylated proteins and a number of minor bands were detected in the post-nuclear supernatant fraction. Only one of these proteins, a prominent band of approximately 42 kDa, exhibited an increased amount of phosphotyrosine when cells were incubated in hypotonic medium for 4 hrs and 10 min vs. isotonic controls which showed no change (Fig. 4). Hypotonic incubation in the presence of the tyrosine kinase inhibitors, genistein ($100\mu M$) and lavendustin B ($50\ \mu M$), significantly inhibited taurine efflux and concomitantly reduced the phosphorylation of the 42 kDa protein, suggesting that this protein is involved in RVD.

These observations suggest that the membrane transporter involved in RVD in dogfish rectal gland cells differs from transporters previously described in skate erythrocytes and hepatocytes, since taurine transport is not inhibited by pyridoxal phosphate, DIDS or NPPB. We also provide evidence that a tyrosine phosphoprotein may be involved.

This work was supported by a New Investigator Award to JKH from the MDIBL. DS was a recipient of a fellowship provided by NSF REU Grant number 93-22221 to MDIBL, and RL and JF were supported by NIH grant numbers DK-34208 and P30ES3828 (Center for Membrane Toxicology Studies).

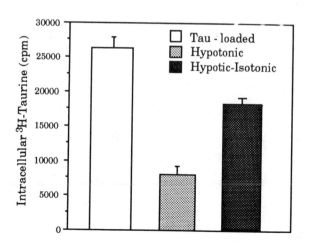

Fig. 1. Effect of medium osmolarity on taurine efflux from shark rectal gland cells (one day old cultures). Results are shown from a representative experiment. Counts per minute (cpm) of ^{3}H-taurine were measured in cells after preloading for 2h and after 15 min incubation in efflux medium. Tau-loaded, cpm after preloading; Hypotonic, cpm after efflux in hypotonic medium; and Hypotonic—Isotonic, cpm after efflux for 5 min in hypotonic medium followed by 10 min in isotonic medium. Values are means ±SE (n=3). All differences are significant (ANOVA, F = 134, p<0.0001 for every Scheffé comparison).

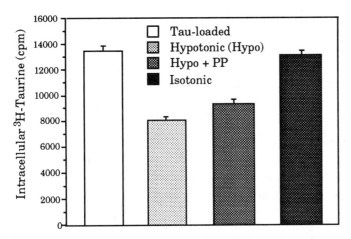

Fig. 2. Effect of pyridoxal phosphate(PP) on taurine efflux from
shark rectal gland cells (2 day old cultures) in hypotonic
medium (600mOsM). Results shown are from a representative
experiment. The same experimental design was used as in Fig.
1. Tau-loaded, cpm after preloading; Hypo + PP, same as Hypo
except efflux medium contained 2mM pyridoxal phosphate;
Isotonic, cpm after efflux in hypotonic medium. Values are
means ±SE (n=3). There is a significant difference between the
mean values of Tau loaded cells and cells after efflux incubation
in hypotonic medium (ANOVA, F = 85, p<0.0001 for Scheffé
comparison). However, there is no significant difference
between the mean values of cells after efflux in hypotonic
medium and hypotonic medium plus pyridoxal phosphate
(ANOVA, p>0.05 for Scheffé comparison).

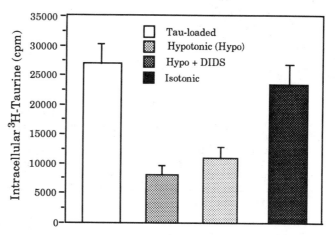

Fig.3. Effect of 4,4'-diisothiocyanostilbene-2,2'- disulfonic acid
(DIDS) on shark rectal gland cells (2 day old cultures) in
hypotonic medium (600mOsM). Results shown are from a
representative experiment. Same experimental design as in
Fig.1. Hypo + DIDS, hypotonic medium contained 0.5μM DIDS.
Values are means ±SE(n=3). There are significant differences
between the mean cpm of Tau-loaded cells and the cells
incubated in hypotonic medium or cells incubated in hypotonic
medium plus DIDS (ANOVA, F=11.3, p<0.0001). There is no
significant difference between the mean cpm of cells incubated
in hypotonic medium and cells incubated in hypotonic medium
plus DIDS (ANOVA, p>0.05).

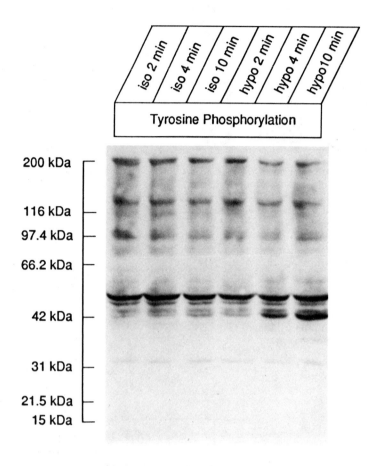

Fig.4. Western blot analysis of tyrosine-phosphorylated peptides in confluent primary cultures of SRG cells incubated in isotonic (control) and hypotonic media for 2-10 min. Postnuclear lysates were generated and subjected to SDS-polyacrylamide gel electrophoresis (7.5% acrylamide) and electrotransferred to membranes. Companion lanes probed with the anti-phosphotyrosine antibody 4G10 revealed multiple bands with a prominent new 42KDa band appearing at 4 minutes in hypotonic but not isotonic media.

METABOLIC FUEL PREFERENCES IN GILL AND LIVER TISSUES FROM FRESHWATER- AND SEAWATER-ACCLIMATED EEL (ANGUILLA ROSTRATA)

Elizabeth L. Crockett[1], Sarah M. Vekasi[2], and Erin E. Wilkes[3]
[1] Department of Biological Sciences, Ohio University, Athens, OH 45701
[2] Mount Desert Island High School, Mount Desert, ME 04660
[3] Ellsworth High School, Ellsworth, ME 04605

Patterns of metabolic fuel utilization may be altered during specific behavioral activities (e.g., sustained locomotion, migration) or by environmental change (e.g., food availability, temperature). Changes in diet (quantity and quality) are likely to occur for migratory animals, and many migrators must rely on stored energy reserves during this period. For the catadromous teleost Anguilla rostrata, adaptation to seawater (and the preparation to move to seawater) may signal a change in metabolic status that could elicit a switch in preferred metabolic fuels. Egginton (J. Exp. Zool. 237: 173-184, 1986) has shown that in aerobic locomotory muscle, capacities for glucose utilization (indicated by hexokinase activities) as well as fatty acid oxidation (indicated by carnitine palmitoyltransferase activities) are higher in silver (freshwater - migratory) eels compared with yellow (freshwater - nonmigratory) eels. In addition, metabolic fuel selectivity in aerobic locomotory muscle from silver eels appears to favor the oxidation of lipid fuels compared to carbohydrates (Egginton, 1986).

While much attention has been devoted to examining metabolic fuel utilization in muscular tissues (skeletal, cardiac), little is known about metabolic fuel preferences of gill epithelia (Mommsen, In Fish Physiology, v 10. eds. WS Hoar and DJ Randall. Academic Press, Orlando. pp. 203-238, 1984). It seems quite plausible that the metabolic fuel preference of gill may change as the teleost prepares for migration and its transition to the marine environment. Like muscle, gill from A. rostrata may have to rely to a greater extent on storage lipids rather than carbohydrate fuels during its migratory phase. Also, mitochondrial densities of gill chloride cells are elevated in teleosts living in seawater compared with animals raised in lower strength seawater (King et al., Cell Tissue Res. 257: 367-377, 1989). Proliferation of mitochondria in gills from seawater animals may also increase the likelihood that lipids become more important fuels during both the transition to seawater and migration. We have determined capacities for glucose and fatty acid oxidation for gill tissues from freshwater- and seawater-acclimated eels. We have also performed similar experiments with liver tissue to compare with our results for gill. We report maximal ATP yields for the two respective metabolic pathways (glycolysis and mitochondrial ß–oxidation).

Yellow eels were captured in freshwater (Penobscot River, Maine) and acclimated to freshwater (recirculating well water with daily turnover) or seawater (flow-through) for a minimum of three weeks prior to use. Temperatures in freshwater aquaria were matched (within 1°C) to seawater temperatures using chilling units. Animals were anesthetized (0.1% neutralized MS-222 dissolved in either freshwater or seawater depending on acclimation group) and gills were perfused with heparinized saline. Livers were removed and gill tissue was scraped from gill arches on an ice-cold glass stage. Liver (20% w/v) and gill (33% w/v) tissues were homogenized in 1 ml Ten Broeck ground glass homogenizers. Enzyme activities were determined using a Beckman DU 640 spectrophotometer. Assays were conducted at ambient temperatures (28°C $\pm$ 1°C). Hexokinase (HK) and carnitine palmitoyltransferase (CPT) activities were measured as the reduction of $NADP^+$ (coupled to glucose-6-phosphate dehydrogenase), and carnitine-dependent reduction of DTNB (by free CoA), respectively. CPT activities were measured with a monounsaturated substrate, palmitoleoyl CoA. While a range of fatty acids may be employed for the CPT assays, it is likely that the monounsaturated substrate will provide activities of CPT that are maximal (Egginton, 1986; Crockett and Sidell, Biochem. J. 289: 427-433, 1993). Since CPT occurs as two forms (CPT I and CPT II), and only CPT I is considered rate-limiting, we estimated CPT I as half of the total CPT activity. ATP yields were calculated as 30 μmoles ATP/μmole glucose and 105 μmoles ATP/μmole palmitoleoyl CoA.

Capacities for ATP generation are 8-9 times greater for glucose than for the lipid fuel, palmitoleoyl CoA, in gills from either freshwater- or seawater-acclimated eels (Figure 1). In contrast with gill, liver tissues can generate at least 4-fold more ATP from oxidation of the lipid fuel compared with ATP yields calculated for carbohydrate utilization. CPT activity is 2-fold greater in hepatic tissues from seawater-acclimated eels than from freshwater-acclimated animals (data not shown). While our small sample size precludes a statistically significant result, the trend implies that there is an even greater capacity for ATP production from the lipid compared with the carbohydrate fuel in liver from seawater-acclimated animals (Figure 1).

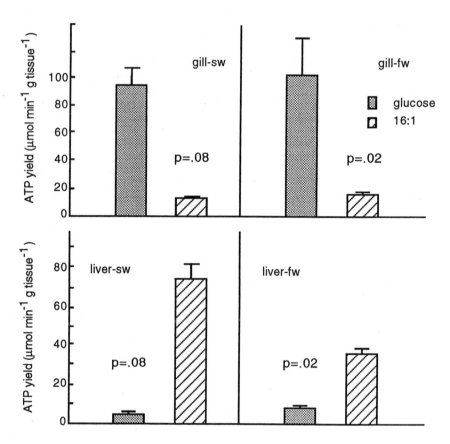

Figure 1. Capacities for ATP generation from the oxidation of glucose or palmitoleoyl CoA (16:1). Top panel: gill from seawater-acclimated (sw) and freshwater-acclimated (fw) animals. Bottom panel: liver from seawater-acclimated (sw) and freshwater-acclimated (fw) fish. Values represent mean ± SEM (sample sizes range from 2 to 4).

Our results suggest that gill tissue is largely reliant on glucose oxidation in freshwater. While a similar result emerges from our study with seawater-acclimated animals, a larger sample size is required to confirm this trend. In marked contrast, hepatic tissues show preference for the oxidation of fatty acid fuels. In addition, the capacity for fatty acid oxidation in liver is increased 2-fold in seawater-acclimated animals compared with freshwater counterparts.

Research supported by a New Investigator Award from MDIBL (Blum/Halsey Scholar Award) and start-up funds from Ohio University to ELC. SMV and EEW were recipients of fellowships from the Hancock County Scholars Program funded by NSF ESI-9452682.

THE CARDIAC EFFECTS OF COPPER AND CERULOPLASMIN ON SPINY DOGFISH SHARK (SQUALUS ACANTHIAS)

Rui Wang[1], Lingyun Wu[1], Yi Zhang[1], Dino Stea[2],
Mircea A. Mateescu[2], Alfred D. Doyle, and McErney R. Branch Jr.
[1]Department of Physiology, Université de Montréal, Montréal, Québec, Canada H3C 3J7;
[2]Department of Chemistry,Université du Québec à Montréal,Montréal,Québec,Canada H3C 3P8

Copper, an essential trace element, participates in maintaining the physiological functions of many organisms. The neuronal effect of copper as well as the interaction of copper with neuronal ion channels have been noticed. The toxicity of copper on the cardiovascular system has not been explored except for the study by Evans & Weingarten (Toxicol. 61:275, 1990) indicating that the contraction of aortic smooth muscle from Squalus acanthias might not be affected by copper. The physiological range of serum copper in dogfish is currently unknown. Over 90% of copper in mammalian serum is bound to ceruloplasmin (CP), a multifunctional blue copper-protein (132 kDa) containing 5-6 copper atoms per molecule. Recently, we showed that CP depolarized neuronal membrane and suppressed neuronal K channel currents (Wang et al., Biochem. Biophys. Res. Commun. 207:599,1995). Since copper and CP are closely related to each other and they may actively participate in the regulation of cardiovascular functions under toxicological or (patho)physiological conditions, the effects of copper and CP on heart beating and on K channel currents in single ventricular myocytes from dogfish shark Squalus acanthias were examined in this study.

In the first series of experiments, the beating rate of atria isolated from pithed Squalus acanthias (male, 2-7 kg) was measured. After removal from the shark, the hearts were immediately immersed in cold elasmobranch physiological saline (EPS) containing (mM): NaCl 270, urea 350, KCl 4, $MgCl_2$ 3, HEPES 10, $CaCl_2$ 3, Na_2SO_4 0.5, KH_2PO_4 0.5, and heparin (50 units/ml). Both aorta and two coronary arteries were cannulated and the whole heart was mounted on a Langendorff device. The time from the removal of the hearts from shark to the initiation of retro-perfusion of isolated heart with oxygenated EPS (30ºC) was less than 30 min. The atrial beating was counted if the beating rate was stable for 15 min. The stable basal beating rate of

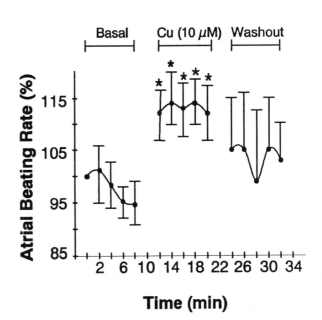

Fig. 1. Effect of copper on atrial beating rate.
* p<0.05 vs. basal beating rate. n= 16.

atria was 49±12 times/min (X±SEM, n=16). $CuSO_4$ increased the atria beating rate at different concentrations (1-100 µM) (Fig. 1) although the dose-response relationship needs to be further clarified. The increased atrium beating rate gradually slowed down toward the basal level after removing $CuSO_4$ from perfusing solution. It was also found that, in some cases, when atrium beating rate increased, ventricle beating rate decreased or even stopped. After the increased atrium beating rate was corrected, ventricle beating rate also recovered and synchronized with atrial beating rate. To test the effect of CP on atrial beating rate, highly purified bovine CP from our laboratory (Wang et al., Prep. Biochem. 24:237,1994) was used. At concentrations of 0.1-10 μM, CP had no effect on atrial beating rate (n=6).

In the second series of experiments, K^+ channel currents of single shark ventricular myocytes were investigated. The similar cell isolation procedure as described by Mitra & Morad (J. Physiol. 457:627, 1992) was followed with modifications. Briefly, after the isolated hearts were cannulated and mounted on a Langendorff device, they were perfused with oxygenated EPS with or without calcium for different periods. Collagenase (0.06%) and hyaluronidase (0.06%) were then added

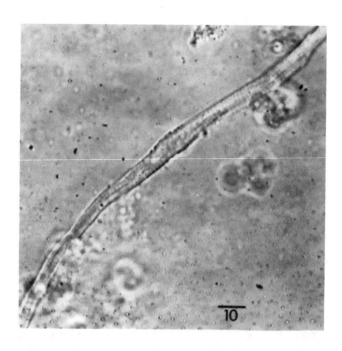

Fig. 2. The morphology of freshly isolated ventricular myocytes from shark heart. The horizontal bar represents 10 μm.

to EPS and the heart was digested for approximately 30 min. Thereafter, the heart was removed from the Langendorff device and atrium and connective tissues were removed. Gentle agitation released single myocytes from the remaining ventricular tissue. The isolated myocytes were characterized by their striated appearance, spindle shape (Fig. 2), and the contraction in response to KCl stimulation. The myocytes were kept at 4°C and used 8-24 hours after isolation for recording the delayed rectifier outward K^+ channel currents with the whole-cell patch-clamp technique. The pipette solution contains (mM): KCl 200, urea 300, trimethylamine N-oxide (TMAO) 150, NaCl 30, HEPES 20, and MgATP 2. The bath solution contains (mM): Trizma 300, KCl 5, $MgCl_2$ 3, KH_2PO_4 0.5, urea 300, HEPES 10, TMAO 70, $CaCl_2$ 0.2, TTX 1 μM. The pH and osmolality of all solutions were adjusted to 7.4 and 920 mOsm/l, respectively. $CuSO_4$ (10 μM) had no effect on K^+ channel currents (n=4). One example was shown in Fig. 3. However, CP (1 μM)

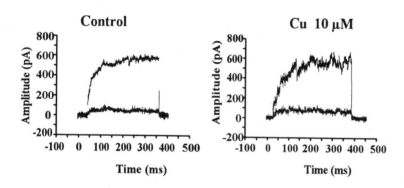

Fig. 3. Effect of copper on K channel currents in one shark ventricular myocyte. The currents were elicited from a holding potential of -80 mV to +10 and +30 mV, respectively.

significantly inhibited K$^+$ channel currents by 53±6% at +20 mV (n=4, p<0.05). This effect of CP was more pronounced at more positive membrane potentials (Fig. 4).

Our results suggest that the cardiac effects of copper and CP have different profiles. Since only atrial beating rate was measured in the present study, effects of copper and CP on the contractility of atrial and/or ventricular muscles are still unknown. The decreased atrial beating rate in the presence of copper, per se, may significantly affect the pump function of shark heart. That may form the basis of cardiac toxicity of copper in this species. CP-induced inhibition of outward K$^+$ channel currents in shark ventricular myocytes was not mediated by copper since this heavy metal did not affect the same ion channels. Although CP had been localized in the plasma of some fishes, such as trout (Perrier et al., Comp. Biochem. Physiol. 49B:679, 1974) and we also found the oxidative activity of shark plasma representative of CP functions, we failed to purify blue CP from shark plasma using our established purification method. This uncertainty of the presence of CP in dogfish shark plasma raised two possibilities. First, our chromatographic method for purification of bovine serum CP (very sensitive to the ionic strength) may be influenced by the unique presence of a great amount of urea in shark plasma. Second, there may be no CP in shark plasma. In the latter case, the modification of K$^+$ channel currents of shark ventricular myocytes by CP may indicate that the K$^+$ channels in shark ventricular myocytes and their mammalian counterparts share the same modification sites or mechanisms to CP. In the future we plan to study the following questions. (1) Do copper and/or CP affect the contraction force of shark cardiac muscles? (2) Does the cardiac effect of copper depend on the valency state of this ion (Tan & Roth, Neuropharmacol. 23(6):683, 1984)? (3) Is the decreased heart beating rate of shark by copper due to the inhibition of calcium channels? (4) Can the action potential and resting potential of shark ventricular myocytes be altered by copper and/or CP? (5) Are there blue CP or other CP-like proteins in shark plasma? (6) Is there any toxic effect of copper on the metabolism and proliferation of ventricular myocytes of Squalus acanthias?

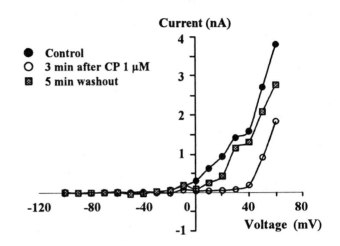

Fig. 4. Effect of CP on K channel currents in one shark ventricular myocyte.

The authors thank Dr. David H. Evans for his constructive advice and sharing his knowledge and experience on cardiovascular physiology of shark. Dr. David S. Miller generously helped us in preparation of photographs of isolated shark ventricular myocytes. This study was supported by a New Investigator Award from MDIBL and CMTS (ES03828-10) to R.W., and by studentship awards from NSF ESI-9452682 to A.D.D. and M.R.B.

EFFECT OF CADMIUM ON ACTIN CYTOSKELETAL STRUCTURE IN SKATE
(RAJA ERINACEA) HEPATOCYTES

Carol E. Semrad[1], David Seward[2], and Alistair Donald[3]
[1]Department of Medicine, Columbia Univeristy College of Physicians and Surgeons,
New York, NY 10032
[2]Summer High School, Sullivan, Maine 04607
[3]Martin Ryan Marine Science Institute, Galway, Ireland

Cadmium (Cd^{2+}) is a heavy metal pollutant of the environment which progressively accumulates in the liver, kidneys and reproductive tissues of mammals and marine animals. The mechanism by which Cd^{2+} damages tissues is unclear. One of the earliest detectable effects of cadmium is an increase in permeability to low molecular weight proteins across mammalian epithelia. Prior studies suggest that this is due to the disruption of tight junctions between epithelial cells. The purpose of this study was to examine the effects of Cd^{2+} on cytoskeletal organization and tight junction permeability in polarized skate hepatocytes.

Hepatocytes from the little skate (Raja erinacea) were isolated by the method described by Smith and Boyer (J. Exp. Zool. 241:291, 1987), resuspended in elasmobranch Ringer's solution and maintained at 15°C. Immunocytochemical studies were done to study the effects of Cd^{2+} on the structural arrangement of F-actin, α-actinin and zona occludens-one (ZO-1), all important in tight junction formation. For these studies, hepatocytes were treated with $CdCl_2$ (100μM, 300μM, 600μM and 1mM) for 30-90min, then allowed to adhere to poly-L-lysine coated coverslips. Cells were fixed in 100% acetone and treated with rhodamine phalloidin for the staining of F-actin, and rat anti-ZO-1 polyclonal antibodies or bovine anti-α-actinin polyclonal antibodies, followed by fluorescein conjugated secondary antibodies, for the staining of ZO-1 and α-actinin. The cells were viewed with a Nikon Microphot 5-A epifluorescent microscope (60X objective, N.A. 1.4). In control (untreated) hepatocytes, actin filaments were present in the pericanalicular and submembranous cortical regions of the cell as described previously by Henson et al. (J. Exp. Zool. 271:273, 1995). α-actinin was present predominantly in the pericanalicular region of hepatocytes. ZO-1 staining was faint and not found in all hepatocyte clusters studied. Different fixation methods did not improve ZO-1 staining. No changes in actin or α-actinin organization were detected in skate hepatocytes treated with $CdCl_2$ (100μM over 90min and up to 600μM over 30 min). Hepatocytes exposed to 1 mM $CdCl_2$ showed both an enlargement in canalicular lumen size and an irregular (ruffled) cortical staining pattern of actin when compared to control cells. This may have been due to cell damage, as confocal fluorescence microscopic studies showed that exposure of hepatocytes to 1mM $CdCl_2$ caused a rapid uptake of fluorescent dextrans into the hepatocyte cytoplasm.

Studies were started to examine the effects of Cd^{2+} on tight junction permeability by measuring the uptake of fluorescent dextrans into the canalicular lumen of hepatocyte clusters using confocal fluorescence microscopy (Meridian Insight Plus frame scanning confocal microscope). Skate hepatocyte clusters with lumens were difficult to identify because of the small lumen diameter. For this reason, hepatocytes were treated with PMA, a phorbol ester which activates PKC and has been shown to increase tight junction permeability in rat hepatocyte clusters (Nathanson et al., Am. J. Physiol., 262:G1079, 1992 and 269:G789, 1995). Treatment of hepatocytes with PMA (1μM) in the presence of fluorescein-dextran (3000 M.W.) resulted in the appearance of dye in the canalicular lumen. Further studies need to be done to determine whether this was due to dye movement through the paracellular space or by transcytosis.

In conclusion, preliminary studies suggest that acute exposure of skate hepatocytes to high concentrations of Cd^{2+} does not alter skate hepatocyte cytoskeletal structure. In contrast, cultured mammalian epithelial cell monolayers develop a decrease in transepithelial resistance (Janecki et al., Tox.

Appl. Pharm., 112:51, 1992) and alterations in F-actin organization (Prozialeck et al., Tox. Appl. Pharm., 107:81, 1991) when exposed to much lower concentrations of Cd^{2+} (10-50μM), albeit over a longer exposure time. Whether Cd^{2+} can alter tight junction permeability without altering cytoskeletal structure in skate hepatocytes, perhaps by interfering with Ca^{2+}-dependent intercellular bridge formation, has not been determined. These findings suggest that skate hepatocytes, unlike mammalian epithelia, are relatively resistant to the effects of Cd^{2+} on cytoskeletal structure. Skate hepatocytes have also been shown to be relatively resistant to the effects of Cd^{2+} on cellular metabolism (Ballatori et al., Tox. Appl. Pharm., 95:279, 1988). The effect of longer exposure times of Cd^{2+} on the cytoskeletal structure of skate hepatocytes needs to be evaluated.

Acknowledgements: Thanks to Drs. Nathanson, Henson, Boyer, and Ballatori for their helpful discussions and Drs. Henson and Soroka for technical assistance. This work was supported by a Young Investigator Award from the Center of Membrane Toxicity Studies (ES03828-10) and NSF ESI-9452682

ROLE OF METALLOPROTEINASES IN MEDIATING HEAVY METAL TOXICITY IN THE DOGFISH SHARK, SQUALUS ACANTHIAS

Matthew Mulvey[1] and Nancy Berliner[2]
[1]Wesleyan University, Middletown, CT
[2]Dept of Medicine and Genetics, Yale University School of Medicine
New Haven, CT

The matrix metalloproteinases (MMPs) are a closely related family of matrix modifying enzymes thought to mediate tissue breakdown at the sites of injury and repair. MMPs play an important role in mediating the inflammatory response as well as being hypothesized to be important in tissue invasion and metastasis (Matrisian L.M., Trends in Genetics 6: 121-125, 1990; Stetler-Stevenson W.G. et al, Ann. Rev. Cell Biol. 9: 541-73, 1993). Although a role has been hypothesized for these enzymes in renal (Davies M. et al, Kidney International 41: 671-678, 1992) and hepatic (Arthur M.J.P., Seminars in Liver Disease 10: 47-55, 1990) lesions associated with cellular proliferation and fibrosis, surprisingly little data have been obtained regarding the modulation of their activity in the setting of tissue injury.

The activity of MMPs can be postulated to be potentially modulated by heavy metals at both the transcriptional and post-translational level. The transcriptional regulatory pathways of metallothionein and the MMPs show considerable overlap, and the two have been shown to be coordinately regulated in chondrocytes in osteoarthritis (Zafarullah M., et al., FEBS 306: 169-172, 1992). This suggests that heavy metals, which induce metallothionein expression, could also effect MMP gene transcription. Furthermore, collagenases are activated in vitro by organomercurials (Mookhtiar K.A. et al., Analytical Biochemistry 158: 322-333, 1986), indicating that interaction of other metal ions with the zinc within the MMPs can modulate enzymatic activity.

In view of the evidence for the potential interaction between heavy metals and pathways of regulation of MMP expression and activity, we undertook to examine the role these enzymes may play in the tissue response to heavy metal poisoning. We have begun to isolate cDNA clones for species-specific MMPs from the dogfish shark rectal gland using degenerate oligonucleotide primers directed against highly conserved regions shared among the members of the MMP family. We plan to use these cDNA probes to analyze the pattern of mRNA expression of the MMPs in a range of normal tissues. Studies will then be undertaken to determine the activity of these genes in response to exposure to heavy metals (mercury, cadmium, nickel) in cultured shark rectal gland cells and in isolated perfused shark rectal glands.

Cloning of the species-specific MMPs was initiated by exploiting the high degree of conservation between two specific regions shared by the whole family of enzymes, namely the "cysteine switch" region and the zinc binding site. These sequences are highly conserved evolutionarily, and have been documented to show homology between mammals and animals as distantly related as the sea urchin (Lepage T., Gache C., EMBO 9: 3003, 1990). An aliquot of the dogfish shark rectal gland library was heated to 95° C to disrupt the phage coat and inactivate endogenous proteases. Subsequently, primers, nucleotides, buffer, and Taq polymerase were added and a low-stringency PCR performed for thirty cycles. Because gelatinases contain a unique fibronectin binding domain absent from collagenases, these two groups of enzymes give a characteristic pattern on PCR, yielding two species of approximately 800bp and 300bp respectively.

PCR was undertaken over a range of temperature and Mg concentrations, and PCR products of the predicted size (300bp and 800bp) were seen. The PCR products were directly cloned into the TA vector system (Stratagene), and restriction digest confirmed inserts of 300 and 800 bp. These plasmids are currently being sequenced to confirm their identity as MMPs. DNA inserts with intact open reading frames and homology to MMPs will be used to screen the parent library to allow us to obtain full-length clones of the cDNAs. These will be further analyzed for their homology to known MMPs from higher species. Although fragments of both predicted sizes (300bp and 800bp) have been obtained, the identity of these fragments must be established by sequencing, as it is quite common to obtain unrelated fragments of coincidentally expected size by low stringency PCR.

It is our hypothesis that study of the pattern of MMP expression in response to heavy metals could provide important insights into the pathogenesis of clinical syndromes associated with toxic metal exposure. Furthermore, it is possible that relevant studies of metalloproteinase expression in response to heavy metals could provide a general model for tissue injury applicable to a wide range of lesions. There is intense interest in the pharmaceutical industry in developing specific inhibitors to metalloproteinases. If it could be established that the pathogenesis of tissue injury involves overexpression of MMPs, this could broaden the potential applicability of such agents.

We are currently sequencing the DNA inserts cloned from the PCR products last summer. Once an insert with specific MMP sequences is identified, this will be used to clone a cDNA clone from the rectal gland library. At the same time, the insert can be used directly as a probe to examine the mRNA expression of the identified gene in shark rectal glands isolated from animals perfused with heavy metals. These results will be correlated with examination of histologic changes induced by heavy metal perfusion.

FUNDING: These studies were supported by a New Investigator Grant from MDIBL. M. Mulvey received a Summer Student Research Fellowship from the American Heart Association, Connecticut Affiliate.

ACUTE EFFECTS OF CERTAIN HEAVY METALS ON <u>MYTILUS EDULUS</u>

George W. Kidder III[1] and Pamela J. Foster[2]
[1]Dept. of Biological Sciences, Illinois State Univ., Normal, IL 61761
[2]John Bapst Memorial High School, Bangor, ME

We (Kidder & McCoy, Bull. MDIBL, 34:102, 1995) have shown that the electromyogram (EMG) is a rapid and sensitive indicator of acute mercury toxicity in the blue mussel. The present study extends these observations to additional heavy metal cations Cd^{2+}, Zn^{2+}, Ni^{2+} and Cr^{3+}. The methods were those previously employed. Briefly, two electrodes (Ag|AgCl, 0.3 mm thick) were inserted between the valves of a 1-2 g mussel secured with cyanoacrylate glue in 25 ml of artificial sea water (ASW) at 12°C. The amplified (gain 2000, 3 db bandwidth 0.3 - 30 Hz) EMG is sampled at 50 Hz, recording the absolute maximum value obtained during each second. The chloride of each cation was dissolved as a 10 mM solution in ASW, and added to the bath to obtain a final concentration of 10 or 100 μM. An attempt to investigate Cu^{2+} was abandoned due to the formation of a precipitate in the stock solution, probably $CuCO_3 \cdot Cu(OH)_2$ (malachite).

Each experiment started with a 30 minutes control period in ASW. The test solution was then added at 10 μM for 30 minutes, and removed by washing (2 changes) with ASW. After an hour in ASW, the test solution was added at 100 μM for 30 minutes, followed by an additional hour of washout. When recording was terminated, the mussel was placed in natural sea water and monitored for at least one week for delayed toxic effects. The 12,600 data points for each experiment were averaged into 5 minute periods, normalized to the average value during the first control period, and these averages (± SE) were collected and graphed as shown in the figure. The number of mussels was: Cd^{2+}, 8; Zn^{2+}, 6; Ni^{2+}, 7; and Cr^{3+}, 6.

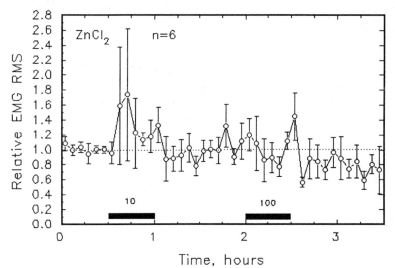

Inspection of the figure shows that while there is a suggestion of a response to $ZnCl_2$, these responses are not statistically significant. A similar lack of response was noted to each of the other cations tested; in no series was statistical significance (5% probability) reached by more than one point in 20. This is in sharp contrast to the previously reported action of mercury, which was significant at 1 μM. Likewise, the longevity of these mussels was not affected by these brief heavy metal exposures, since there were no deaths observed during the week of post-experimental observation.

We must conclude that whatever the chronic effects of these ions may be, there is no demonstrable effect of short term exposure to these metals on either the EMG or the one-week survival.

Supported in part by NSF-ESI-9452682.

METALLOTHIONEIN INDUCTION DURING SPERMATOGENESIS IN SQUALUS ACANTHIAS

Marlies Betka
Department of Biology
Boston University, Boston, MA 02215

Epidemiological studies indicate that spermatogenesis may be threatened by environmental pollutants (decline in sperm count from 133 million/ml in 1940's to 60 million/ml in 1990's; Carlsen et al., British Med. J. 305:6854, 1992). Cadmium (Cd), although found only in trace concentrations in the environment, has a long biological half life (10-30 yr in humans) and thus may be a cumulative toxicant (Coopius-Peereboom & Coopius-Peereboom-Stegeman, Tox. Environm. Chem. Rev. 4:67-178, 1981). In the mammalian testis, Cd is a potent spermato-toxicant with damaging effects at doses well below those that induce kidney damage (Singhai et al., In: Thomas JA (Ed): Endocrine Toxicology, 149-179, 1985). The testis of the spiny dogfish shark (Squalus acanthias) has been shown to have a Cd accumulating mechanism (Betka & Callard, unpublished). Tissues can accumulate Cd by metallothionein (Mt) binding. Mts are a family of low-molecular-weight metal-binding proteins. Kille et al. (Biochim. Biophys. Acta 1089:407-410, 1991) took advantage of high sequence homology in the 5'end of various Mt cDNAs and have successfully used PCR techniques to isolate specific Mt RNA transcripts in a variety of freshwater and saltwater fish. This approach was also used to isolate Mt cDNA from shark testis to gain a molecular probe to investigate the role of this protein in Cd accumulation.

Three sharks were injected with a single dose of CdCl$_2$ (5mg/kg body weight) into the caudal vein and sacrificed after 3 days. Testis of treated and control animals were collected and dissected into premeiotic (PrM), meiotic (M) and postmeiotic (PoM) stages. Total RNA was extracted by the method of Chomczynski & Saachi (Anal. Biochem. 162(1):156, 1987). Reverse transcription was performed with the SUPERSCRIPT Kit (Gibco BRL) and the resulting cDNAs were PCR amplified. Primers used were as described by Kille et al. (Biochim. Biophys. Acta 1089:407-410, 1991). Bands of the expected size of approximately 320 bp were found in all three stages in Cd-treated animals and in M and PoM tissues of control animals. These fragments were isolated and subcloned (TA Cloning Kit, Invitrogen) and sequencing is in progress. After confirmation of the identity of Mt, these probes will be used in Northern analysis of Cd treatment experiments (in vivo and in vitro).

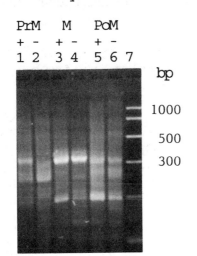

Fig. 1: PCR amplified products in shark testis. The PCR reactions were performed in a Perkin Elmer DNA Thermal Cycler 480 at 94°C (1 min), 55°C (1 min) and 72°C (1 min) (30 cycles) with an extension period at 72°C (15 min) and 25°C (5 min). 10 μl aliquots of a 50 μl reaction were run out on a 2% agarose gel. Lane 1/2 - premeiotic (PrM), lane 3/4 - meiotic (M) and lane 5/6 postmeiotic (PoM) testicular tissue from Cd treated (+) and untreated (-) sharks; lane 7 - PCR standards (1000, 750, 500, 300, 150, 50 bp).

This work was supported by NIEHS P42 ES-07381 and a MDIBL Young Investigator Fellowship.

A PROSTAGLANDIN, NOT NITRIC OXIDE, IS THE ENDOTHELIUM-DERIVED RELAXING FACTOR IN THE SHARK (<u>SQUALUS</u> <u>ACANTHIAS</u>) VENTRAL AORTA

David H. Evans and Mark Gunderson

Department of Zoology, University of Florida, Gainesville, FL 32611

In mammals, the vascular endothelium produces an endothelium-derived relaxing factor (EDRF) when it is stimulated by a variety of mediators, including acetylcholine (ACh) and endothelin (ET). It is now generally accepted that the EDRF is the gas, nitric oxide (NO; e.g., (Vane, Philos Trans R Soc Lond [Biol] 343: 225-246, 1994)). Our finding that both ACh (unpublished results) and endothelin (Evans and Gunderson, J Comp Physiol, in press: 1996) produce contraction in aortic vascular smooth muscle (VSM) rings from the spiny dogfish shark, <u>Squalus</u> <u>acanthias</u>, in the presence of an intact endothelium, suggests that this system may be missing in fish VSM. The fact that neither L-arginine nor sodium nitroprusside (NO precursors) stimulate dilation in shark VSM rings (Evans and Gunderson, Bull. MDIBL 34: 109, 1995) supports this hypothesis. In mammals, the vascular endothelium also produces prostaglandins (PGs), such as prostacyclin (PGI_2) and PGE_1, which are dilatory (Vane, Op. Cit.). PGs, however, are not considered to be major EDRFs in mammals (e.g., (Inagami et al., Annu Rev Physiol 57: 171-189, 1995), largely because the original description of the EDRF showed that its production was not altered by inhibition of prostaglandin synthesis (Furchgott and Zawadzki, Nature 288: 373-376, 1980). Nevertheless, it is possible that PGs, rather than NO, may be the dominant EDRF in fishes, and this has been suggested by Miller and Vanhoutte (in Endothelial Regulation of Vascular Tone, Ryan and Rubanyi, eds, pg. 3-20, 1992), based upon work on the ventral aorta of the rainbow trout, <u>Oncorhynchus</u> <u>mykiss</u>.

To test this hypothesis further, we examined the ability of a known NO synthesis inhibitor, L-NAME, and a known prostaglandin synthesis inhibitor, indomethacin, to inhibit the dilation produced by the Ca^{2+} ionophore A23187 in shark VSM rings with an intact endothelium (Evans and Cegelis, Bull. MDIBL 33: 113, 1994). In addition, we examined the ability of NO, PGI_2, and PGE_1 (agonist of PGE_2) to dilate rings after the endothelium had been removed. Rings were prepared and mounted as described previously (e.g., Evans, J Comp Physiol 162: 179-183, 1992), except that tension was monitored using a 4-channel, Biopac A/D recording system connected to a Macintosh 140 Powerbook. In the first series of experiments, endothelium-intact rings were incubated in either 10^{-4} M L-NAME or 10^{-5} M indomethacin (paired with distilled water controls) for 30 minutes before adding 10^{-5} M A23187. In the second series of experiments, endothelium-free rings were exposed to 10^{-6} M of either PGI_2 or PGE_1, or NO (calculated to be $\approx$ 3 x 10^{-5} M), after an initial equilibration period.

L-NAME did not inhibit the A23187-induced dilation in intact aortic VSM rings from the dogfish shark (N = 6), but indomethacin did, actually reversing the 198 $\pm$ 45 mg dilation to an 145 $\pm$ 52 mg contraction in paired rings (initial tension = 500 mg; N = 6; p < 0.01). Nitric oxide did not produce any dilation in this preparation, although the same NO solution dilated rat thoracic aortae in our control experiments. On the other hand, PGI_2 dilated endothelium-free rings by 95 $\pm$ 48 mg (SE; N = 8) and PGE_1 by 389 $\pm$ 121 mg (N = 8), showing that PGs are potent EDRFs in this system. Thus, we conclude that, as hypothesized by Miller and Vanhoutte (Op. Cit.) for a single species of teleost fish, the aortic vascular endothelium of the spiny dogfish produces an EDRF that is a prostaglandin, not nitric oxide. (Supported by NSF IBN-9306997 and a Grant in Aid from the Maine Affiliate of the American Heart Association to DHE, a fellowship from the AHA to MG, and EHS-P30-ESO3828 to the Center for Membrane Toxicity Studies).

CADMIUM INHIBITS ENDOTHELIN-INDUCED, BUT NOT ACETYLCHOLINE-INDUCED, CONTRACTION OF THE SHARK (<u>SQUALUS ACANTHIAS</u>) VENTRAL AORTIC VASCULAR SMOOTH MUSCLE

David H. Evans and Mark Gunderson
Department of Zoology, University of Florida, Gainesville, FL 32611

We have shown previously that cadmium (Cd^{2+}) produces contraction in isolated rings of vascular smooth muscle (VSM) of the ventral aorta of the spiny dogfish shark, <u>Squalus acanthias</u> (Evans and Weingarten, Toxicology 61: 275-281, 1990) and have hypothesized that the heavy metal interacted with muscarinic receptors because atropine blocked the response (Evans et al., Toxicology 62: 89-94, 1990). We have also recently characterized the receptor involved in the contractile response of the same tissue to the peptide endothelin (ET) (Evans and Gunderson, J Comp Physiol, in press, 1996). Wada et al. (Febs Lett 285: 71-74, 1991) have shown that Cd^{2+} displaces ET from receptors on the human placenta, non-competitively displaces ET from solubilized receptors, and inhibits ET-induced contraction of the rat aorta, suggesting that Cd^{2+} is a non-competitive inhibitor of ET receptors. Interestingly, Smith et al. (Environ Health Perspect 102 Suppl 3: 181-189, 1994) have recently proposed that Cd^{2+} interacts with an orphan receptor (no known ligand) which activates IP_3 formation. This proposed orphan receptor shares many characteristics with ET receptors. Thus, we initiated a series of experiments to determine if Cd^{2+} would interfere with either ACh- or ET-induced contraction of shark aortic VSM.

Rings were prepared and mounted as we have described previously (e.g., Evans, J Comp Physiol 162: 179-183, 1992), except that tension was monitored using a 4-channel, Biopac A/D recording system connected to a Macintosh 140 Powerbook. Rings were paired and, after an initial equilibration period at 500 mg tension, one was treated with Cd^{2+} (10^{-5} or 10^{-4} M) for 30 minutes, the other with the same volume of distilled water (10 or 100 µl in 10 ml elasmobranch Ringer's solution), before the addition of 10^{-4} M Ach or 10^{-7} M ET-1.

Incubation of shark aortic VSM rings in 10^{-4} M Cd^{2+} did not inhibit the contraction produced by subsequent addition of 10^{-4} M Ach ($p > 0.10$; N = 13), which suggests that, contrary to our previous hypothesis, Cd^{2+} may not interact directly with muscarinic receptors. The response to ET, on the other hand, was diminished significantly when the rings were incubated in 10^{-4} M ACh (Fig. 1). However, a lower concentration (10^{-5} M) of ACh was without effect.

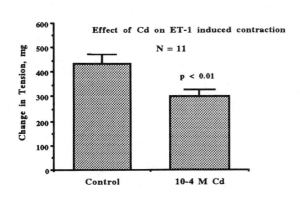

Figure 1

Our data suggest that Cd^{2+} may contract shark aortic VSM by interacting with ET receptors. This may account for the 50% of the Cd^{2+} induced contraction that is insensitive to atropine (Evans et al., Toxicology 62: 89-94, 1990). How Cd^{2+} stimulates atropine-sensitive contraction is still unknown. Supported by NSF IBN-9306997 and a Grant in Aid from the Maine Affiliate of the American Heart Association to DHE, a fellowship from the AHA to MG, and EHS-P30-ESO3828 to the Center for Membrane Toxicity Studies.).

ANALYSIS OF THE APOPTOTIC FORM OF PROGRAMMED CELL DEATH (PCD) DURING SPERMATOGENESIS IN SPINY DOGFISH (SQUALUS ACANTHIAS)

Leon M. McClusky[1], Marlies Betka[1], David Miller[2] and Gloria V. Callard[1]
[1]Department of Biology, Boston University, MA 02215 and
[2]NIH-NIEHS, Research Triangle Park, NC 27709

Due to a cystic mode of spermatogenesis and a simple diametrical arrangement of succeeding germ cell stages in the testis of the shark, this organ is ideal for studying spermatogenesis and its regulation stepwise through development (see review, Callard et al., In: Function of Somatic Cells in the Testis, Bartke A., ed., Springer, NY, p.27-54, 1994). The apoptotic form of programmed cell death (PCD) has been identified as a mechanism for specifically regulating the premeiotic (PrM=stem cell/ spermatogonial) cyst population in shark testis (Callard, et al. Develop. Genetics 16:140-147, 1995). Germ cells but not Sertoli cells are affected all-or-none in a given cyst, but Sertoli cells subsequently phagocytize germ cell remnants. The winter period of spermatogenic inactivity or hypophysectomy induces a zone of degenerating cysts (ZD) between PrM and meiotic (M) stages. Biochemical analysis shows extensive internucleosomal DNA fragmentation (DNA "ladders" on agarose gels), an early marker of PCD, in PrM and ZD stages only, in vivo and in vitro, but this method is not sufficiently sensitive for routine regulatory and toxicological studies. Here we describe a procedure for visualizing PCD in living spermatocysts using acridine orange (AO) as a vital stain. AO has been reported to label condensed chromatin and apoptotic bodies characteristic of cells undergoing PCD during Drosophila embryogenesis, and also labels corpses of apoptotic cells after phagocytosis by macrophages, but necrotic cells are never labeled (Abrams et al. Development 117, 29-43, 1993).

Razor blade-cut cross-sections (≈500 μm thick) of freshly dissected testes or isolated PrM or ZD cysts were cultured in basal medium for 0-7 d as previously described (DuBois and Callard, J. Exp. Zool. 258, 359-372, 1991) with/without various additives. AO (5 μg/ml) was added for the final 30 min of culture. Tissues were washed twice with fresh medium and viewed using an Olympus IMT-2 inverted epifluorescence microscope (502 nm excitation and 526 nm emission maxima) coupled to a Hamamatsu 2400 CCD video camera controlled from a Power Macintosh computer by using the NIH-Image (version 1.57) software package. An AG-5 frame grabber (Scion), which converted the analog video signal into digital values was installed in the computer. Images were captured, averaged, and analyzed or stored for later analysis. In both tissue slices and dispersed spermatocyst cultures, a subset of cysts displayed punctate green or yellow fluorescent spots, as previously described for cells undergoing apoptosis in Drosophila. The presence of apoptotic germ cells in AO-positive cysts was verified by electron microscopy (not shown).

Results indicate that the cysts undergoing apoptosis or nearing the end of the apoptotic process can be identified in both tissue cross-sections and dispersed cyst cultures. Initial data show that the percentage of cysts labeled are affected by stage of development, time in culture, and in vivo or in vitro exposure to added regulators (e.g., steroids) or spermatotoxicants (e.g, cadmium). The technique of AO labeling will allow study of cellular and molecular mechanisms controlling programmed cell death during male germ cell development and is amenable to future analysis by confocal imaging. Supported by grants to GVC (NIEHS P42 ES-07381), LM (Foundation for Research Development, South African Government) and MB (MDIBL Young Investigator Fellowship).

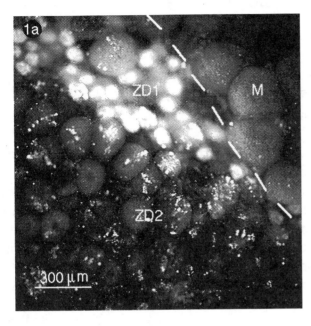

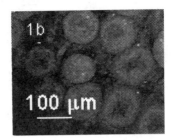

Figure 1. Fluorescence microscopic images of apoptotic cysts in testicular cross section. (a) Spermatogenic development proceeds from lower left to upper right. Note intense labeling of shrunken cysts in ZD_1, indicating an advanced stage of apoptosis and cyst degeneration. Yellow fluorescence suggests AO labels phagocytized germ cell corpses present in Sertoli cells. Late-stage PrM cysts in adjacent ZD_2 had punctate green fluorescent spots, indicating an earlier stage of germ cell apoptosis. Dotted line indicates boundary of PrM(ZD) and M stages. M-stage cysts were never labeled. (b) Small early-stage PrM cysts normally positioned out of the field of view in (a) were mostly unlabeled. Although AO facilitated visualization of cell nuclei in healthy cysts, especially those with mature spermatids undergoing condensation during spermiogenesis, this form of labeling was easily distinguished from labeling of apoptotic cells.

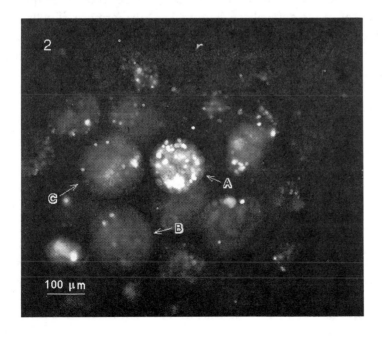

Figure 2. Isolated PrM cysts after 7 days of culture in basal medium. Eighty percent of cysts in this culture were AO-positive. Cyst A which has large yellow spots may represent an advanced stage of apoptosis characteristic of ZD cysts. Cysts B and C have only a few punctate yellow spots and may represent earlier stages of apoptosis.

CELL CYCLE REGULATION IN THE DINOFLAGELLATE, <u>GAMBIERDISCUS TOXICUS</u>: MITOSIS IS COUPLED TO THE DIURNAL CYCLE BY A BLUE LIGHT DEPENDENT SIGNAL

F.M. Van Dolah[1,2,3], T.A. Leighfield[1,3], and L.M. Sugg[3]
[1]Marine Biotoxins Program, U.S. National Marine Fisheries Service,
Charleston Laboratory, Charleston, SC
[2]Medical University of South Carolina, Charleston, SC
[3]University of Charleston, Charleston, SC

Naturally occurring toxins produced by marine microalgae are the major source of food poisonings associated with consumption of seafoods. The frequency, severity and global distribution of blooms of toxic microalgae have increased during the past twenty years. In order to gain understanding of how environmental cues may trigger the rapid growth and reproduction of dinoflagellates that constitutes a "bloom", we have undertaken the current project to identify molecular mechanisms which regulate the cell division cycle in dinoflagellates. Our current work focuses on <u>Gambierdiscus toxicus</u>, a dinoflagellate implicated to be the primary source of toxins responsible for ciguatera fish poisoning. Ciguatera is a potentially fatal neurological syndrome associated with tropical reefs, which afflicts more than 50,000 people annually on a world wide basis.

We have previously reported that cell division in <u>Gambierdiscus toxicus</u> is phased to the diurnal cycle, such that cells divide only during a three hour window late in the dark phase when grown in a 16:8 hour light:dark (L:D) cycle (Van Dolah et al., J. Phycol. 31:395-400, 1995). Formally, cell division phased to the diurnal cycle may be regulated by either the D:L transition (e.g., mitosis occurs 22 h after the onset of light) or the L:D transition (e.g., mitosis occurs 6 h after the onset of dark). To distinguish which transition is critical to cell cycle control in <u>G</u>. <u>toxicus</u>, cells were transferred from a 16:8 to a 12:12 L:D cycle, and the timing of mitosis determined by mitotic index (% of cells possessing replicating nuclei, determined in propidium iodide stained cells). Under these conditions, mitosis continued to occur 6 h after the onset of the dark phase (Figure 1). Furthermore, if cells are placed in 24 h light, such that the L:D transition does not occur, no cells proceed through the cell cycle to mitosis. This indicates that the L:D transition is involved in phasing cell division to the diurnal cycle.

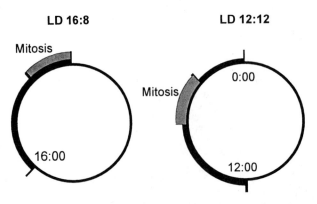

Figure 1. Cell cycle progression is controlled by the L:D transition (Wide bar represents dark; narrow represents light phase).

There are numerous examples of diurnal phasing of cell division among protists as well as metazoans and higher plants (for reviews see Edmunds, Cell Cycle Clocks, Dekker, New York, 1984). However, the mechanisms by which the diurnal cycle entrains the cell cycle remain elusive in all systems examined. Since many metabolic processes in plants are dependent on photosynthetic activity, we first investigated the role of photosynthesis in regulating cell cycle progression in G. toxicus. To determine if the signal permitting cell cycle progression at the L:D transition is coupled to the cessation of photosynthesis, cells at the time of the normal L:D transition were either transferred to dark or maintained in the light, in the absence or presence of the photosystem II inhibitor dichlorodimethylurea (DCMU; 10 μg/ml). Cells transferred to dark entered mitosis in 6 h, as observed previously. By contrast, cells which did not receive the L:D signal did not enter mitosis, regardless of whether photosynthesis continued or was inhibited with DCMU (Figure 2). (DCMU did cause a partial decline in the number dark treated of cells progressing into mitosis, reflecting a toxic effect of long term exposure to DCMU). These results indicate that the signal permitting cell cycle progression through the L:D transition is independent of photosynthesis.

In plants, two signal transduction pathways have been identified which relay light dependent signals to intracellular targets to elicit cellular responses (for review, Short and Briggs, Annu. Rev. Plant Physiol. Mol. Biol 45: 143-71, 1994). One pathway is sensitive to red light (mediated by phytochromes) and one sensitive to blue light (mediated by cryptochromes). We therefore investigated the involvement of red light and blue light in transduction of light dependent signals which may promote cell cycle progression at the L:D transition (Figure 3). At the time of the normal L:D transition, cells were either transferred to dark, maintained in white light, or were exposed to blue light only (Roscolux Filter No. 19) or red light only (Roscolux Filter No. 36). Mitotic index was then determined 6 h later. Cells maintained in light, which did not receive the L:D transition, did not enter mitosis 6 h after the normal time of L:D (or at any time during that diurnal cycle, data not shown). Cells exposed to blue light only (cessation of red light) similarly did not proceed to mitosis. By contrast, cells exposed to red light (cessation of blue light) proceeded into mitosis as if they were in the dark. This suggests that a blue light dependent signal transduction pathway is involved in cell cycle progression at the L:D transition.

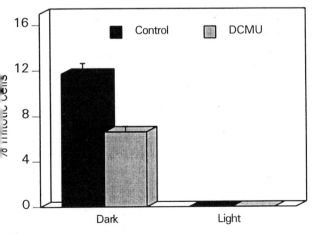

Figure 2. Effect of photosynthesis inhibitor, DCMU, on progression through the L:D transition.

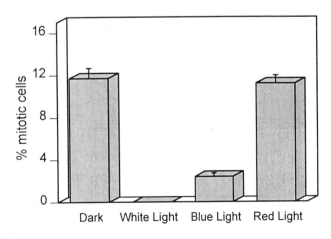

Figure 3. Involvement of red and blue light in cell cycle progression through the L:D transition.

The eukaryotic cell cycle is regulated by two classes of proteins, the cyclins and the cyclin dependent kinases, which together control both the initiation of DNA synthesis and the entry into mitosis (Murray and Kirschner, Science 246:614-21, 1989). We have recently demonstrated the presence of a cdc2-like kinase in G. toxicus, which is activated concurrent with the onset of mitosis, indicating that dinoflagellates possess a similar cell cycle machinery as that present in higher eukaryotes. Cell cycle progression is generally regulated at two control points, the G1/S phase transition and the G2/M phase transition, which coincide with the activation of cyclin dependent kinase complexes. We therefore sought to determine if the L:D transition regulates progression of the G. toxicus cell cycle at G1/S or G2/M. G. toxicus cells were treated at different times prior to mitosis with the S-phase inhibitor, aphidicolin (10µg/ml) and permitted to continue until the normal time of mitosis (22 h after onset of light). The number of mitotic cells present in aphidicolin treated cultures (black bars) at 22 h was then determined relative to untreated controls (grey bar). Aphidicolin added as late as 3 h after the onset of dark (denoted by dark bar on the X-axis) prevented entry into mitosis. This suggests that S-phase is not completed until after the onset of dark. Therefore, the L:D transition most likely controls cell cycle progression prior to the G2/M transition, possibly at the onset of S-phase.

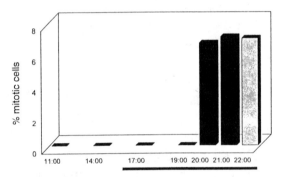

Figure 4. Effect of S-phase inhibitor, aphidicolin, on the percent of cells in mitosis at 22:00 h after onset of light (black bars), relative to untreated controls (grey bar). L:D transition (dark denoted by black horizontal bar) occurred at 16:00 h.

We are currently investigating components of the signal transduction pathway that relays the cessation of blue light to the cell cycle engine. If S-phase entry coincides with the L:D transition, then our results suggest that blue light provides an inhibitory signal which prevents progression of the cell into the cell division cycle. At the G1/S checkpoint in yeast and metazoans, the cell assesses that adequate nutrients are available and that adequate cell size has been attained prior to committing to the energy expensive processes of DNA replication and cell division. Negative regulators of growth (e.g., growth inhibitory polypeptides) have been demonstrated to act at this checkpoint. In G. toxicus, blue light may provide an inhibitory signal which prevents the cell from committing to replication during the light phase of the diurnal cycle.

This work was supported in part by a Milbury Fellowship through MDIBL (FMVD) and by funds from NOAA National Marine Fisheries Service.

IDENTIFICATION OF STIFFENING AND PLASTICIZING FACTORS IN SEA CUCUMBER (CUCUMARIA FRONDOSA) DERMIS

Magdalena M. Koob-Emunds, John A. Trotter[1] and Thomas J. Koob[2]
Mount Desert Island Biological Laboratory, Salsbury Cove, ME 04672
[1]Department of Anatomy, University of New Mexico, Albuquerque, NM 87131
[2]Skeletal Biology Section, Shriners Hospital, Tampa Unit, Tampa, FL 33612

Certain echinoderm connective tissues can rapidly alternate between stiff and compliant states in response to a variety of environmental and mechanical cues (for recent review see Wilkie, in Echinoderm Studies 5, Ed. Jangoux & Lawrence, 1996). Regulation of tissue stiffness in situ is mediated by resident neurosecretory cells through an as yet undetermined mechanism affecting extracellular matrix mechanical properties. Like all echinoderm tissues so far investigated, the dermis of the sea cucumber, Cucumaria frondosa, responds to changes in extracellular calcium, becoming compliant in the absence of calcium and stiff when calcium levels are returned to normal. Previous studies have demonstrated that this experimental modulation of tissue stiffness, results from a positive effect of extracellular Ca^{2+} not on the extracellular matrix directly, but rather on calcium-channel mediated secretion of cytoplasmic granule-bound molecules (Trotter & Koob, J. Exp. Biol. 198, 1951-1961, 1995). In addition, we have presented preliminary evidence for an organic stiffening agent released by cells (op. cit.). The present report describes the identification of this stiffening agent and of an organic plasticizing molecule.

Cucumaria frondosa were obtained from September through November and maintained in the flow-through sea water tanks at MDIBL. All specimens were prepared from the dermis of the two ventral interambulacra that lack podia. The white inner dermis was used to prepare extracts containing stiffening activity, which was bioassayed as previously described (Trotter & Koob, Bull. MDIBL 34, 6-9, 1995). The black outer dermis from which the epidermis had been removed was used to prepare extracts containing plasticizing activity. The principal test solutions used in these studies were MOPS-buffered artificial sea water (ASW), which consisted of 0.5 M NaCl, 0.05 M $MgCl_2$, 0.01 M $CaCl_2$, 0.01 M KCl, and 0.01 M 3-(N-morpholino)propane sulfonic acid (MOPS), pH 8.0; and EGTA-ASW in which the $CaCl_2$ was replaced by 7.2 mM ethyleneglycol-bis-(β-aminoethyl ether) N,N,N',N'-tetraacetic acid (EGTA). As previously established, ASW rendered freshly excised specimens of the inner dermis stiff, whereas these specimens incubated in EGTA-ASW were plastic.

For the preparation of freeze-thaw (FT) extracts, the dermis was cut into roughly 2 x 4 x 6 mm pieces and the dark pigmented outer dermis, including the epidermis, were separated from the white inner dermis with a razor blade. For isolation of stiffening agents, 20g of the inner dermis were finely minced into 1 mm^3 pieces and incubated at ambient sea water temperature in 5 volumes of EGTA-ASW for 5h with intermittent gentle agitation, after which the fluid was decanted and replaced by the same volume of fresh EGTA-ASW. The tissue was frozen in EGTA-ASW for at least 2h at $-70^{\circ}C$, followed by incubation at sea water temperature until it was completely thawed. These steps were repeated for a total of five FT cycles. To prepare extracts of the outer dermis, the epidermis was scraped off with a razor blade and the remaining pigmented

101

part of the outer dermis was minced and extracted with ASW. All other steps were carried out as described above for the inner dermis, except that ASW was used instead of EGTA-ASW for the initial extract. The FT extracts were clarified by centrifugation at 27,000 x g for 30 min. These extracts were then fractionated by anion exchange chromatography, gel filtration chromatography, chromatofusing and hydrophobic interaction chromatography (described below). To assay the extracts and fractions for stiffening and plasticizing activity, four to five replicate specimens were tested in bending. All extracts and fractions to be tested were dialyzed into either EGTA-ASW (stiffener) or ASW (plasticizer) prior to the bending tests.

To prepare uniform test specimens for bending tests, the safety shields of single-edged razor blades were screwed together, either directly or with an intervening metal shim. This cutting apparatus was used to produce equivalent specimens from the white inner dermis that were approximately 3 cm long, 0.85 mm thick, and 1.7 mm wide. The long axis of each specimen was parallel to that of the animal. Gravity bending tests were performed as previously described (Trotter & Koob, Bull. MDIBL op. cit.). The time required for the specimen end to move a vertical distance of 4 mm was measured. All specimens tested were initially incubated for at least 90 min. in either ASW (for plasticizer) or in EGTA-ASW (for stiffener), before they were placed into test solutions. Another 90 min. incubation period in the test solution preceded the bending tests.

Stiffener: FT extracts of the inner dermis were initially applied to a 5 ml anion exchange HiTrap Q column (Pharmacia) in 0.8 M NaCl, 20 mM Tris-HCl, pH 8.0. Stiffening activity was found exclusively in the unbound fraction, whereas all the proteoglycans in the extract were bound to the column, but could be eluted in 3 M NaCl. Following dialysis to reduce NaCl to 0.05 M, the flow through was chromatographed over the same column using a linear gradient from 0.05 to 0.8 M NaCl in 20 mM Tris-HCl, pH 8.0. Bending tests showed that three 2ml fractions eluting between 0.2 and 0.3 M NaCl contained stiffening activity. These fractions were combined, dialysed against 0.8 M NaCl, 20 mM Tris-HCl, pH 8.0 and chromatographed over a 1.6 x 90 cm Sephacryl S-200 gel filtration column (Pharmacia). Bending tests showed that three 5 ml fractions eluting with K_{av} between 0.35 and 0.45 contained stiffening activity (Fig. 1A, fractions 19-21). SDS/PAGE analysis determined that only one protein with apparent molecular weight of approximately 38 kDa was common to these three fractions (Fig. 1B). Chromatofocusing and hydrophobic interaction chromatography confirmed that this protein contained the stiffening activity.

Plasticizer: Only the freeze-thaw extract of the outer dermis contained plasticizing activity; the initial ASW extract had no such activity. This FT extract was subsequently fractionated as described above for the stiffener. Plasticizing activity was found only in the 0.8 M NaCl flow through when first chromatographed on the HiTrap Q column. Only one 2 ml fraction at 0.26 M NaCl from the HiTrap Q displayed plasticizing activity in bending tests. This fraction was subsequently chromatographed on a 1.6 x 60 cm Sephacryl S-100 HR column in 0.8 M NaCl, 20 mM Tris-HCl, pH 8.0; 4 ml fractions were collected. Bending tests determined that one fraction eluting with a K_{av} 0.5 displayed plasticizing activity (Fig. 2, fraction 19). SDS/PAGE analysis of this fraction showed a single Coomassie staining band with an apparent molecular weight of approximately 10 kDa.

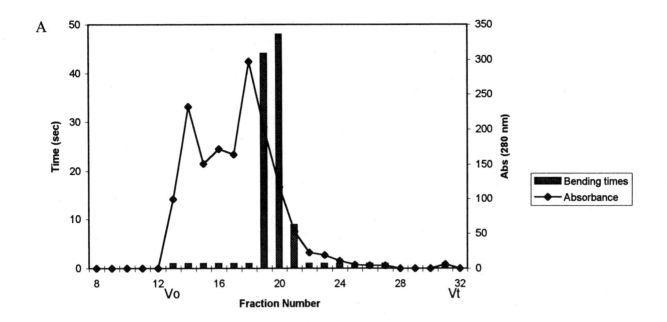

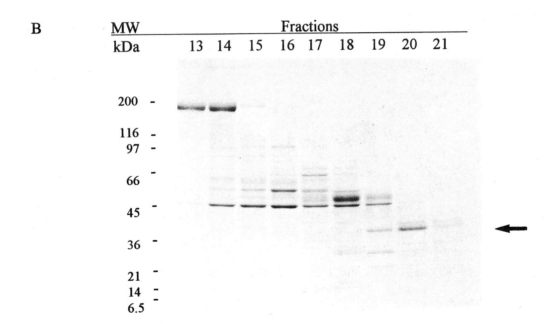

Figure 1. Purification of a stiffening factor from the inner dermis of <u>Cucumaria</u> <u>frondosa</u>. A) Chromatography of active fractions from anion exchange chromatography on Sephacryl S-200 HR. The bars show the results from the bending tests. n = 5 for each bar. B) 4-20% linear gradient SDS/PAGE of fractions from Sephacryl S-200 HR. Arrow indicates the protein with stiffening activity.

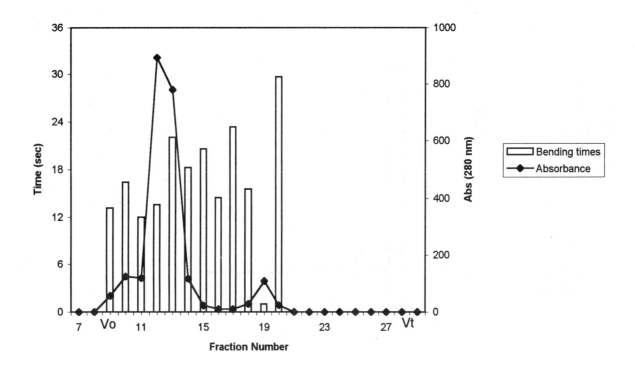

Figure 2. Purification of a plasticizing factor from the outer dermis of <u>Cucumaria frondosa</u>. Chromatography of the active fraction from anion exchange chromatography on Sephacryl S-100 HR. Bars show the results of the bending tests (n = 5 for each bar).

These results confirm and extend our previous observation that modulation of extracellular calcium is not required to alter the compliance of the sea cucumber dermis when live specimens are tested. The purified stiffener increased tissue viscosity in the presence of a calcium chelator (EGTA-ASW). Moreover, the purified plasticizer increased tissue compliance in the presence of normal amounts of calcium (ASW). Since both the stiffener and plasticizer were released only from tissues in which cells were lysed by five freeze-thaw cycles, it appears that significant amounts of these factors are stored within cells. Taken together, these observations suggest that cells in the sea cucumber dermis release either the stiffener or the plasticizer and that these act directly on the matrix macromolecules which mediate tissue mechanical properties.

The identification of these regulatory proteins and development of methods for isolating them in active form, as described here, will allow future experiments to be directed towards: 1) studying the mechanism by which they regulate viscosity of the extracellular matrix of the dermis both in the tissue and with isolated macromolecules; 2) characterizing their biochemical and biophysical properties that are functionally important; 3) delineating the physiological factors that control their secretion.

Funded by grants from ONR and NSF to J.A.T.

BIOMECHANICAL PROPERTIES OF HAGFISH (MYXINE GLUTINOSA) NOTOCHORD

Jan T. Kielstein, Hilmar Stolte[1] and Thomas J. Koob[2]
Mount Desert Island Biological Laboratory, Salsbury Cove, ME 04672
[1]Experimental Nephrology, Department of Internal Medicine, Medical School Hannover, Germany
[2]Skeletal Biology Section, Shriners Hospital, Tampa Unit, Tampa, FL 33612

In most vertebrates, the axial skeleton passes through three phases of development, the first of which is the embryonic notochord. However, in cyclostomes, the notochord remains the sole axial support throughout life. The notochord in adult cyclostomes is a fiber wound cylinder comprised of a relatively thin collagenous sheath enclosing a basement membrane delimited cellular core. Physicochemical tests previously demonstrated that the notochord of Myxine glutinosa is osmotically active and that fixed charge density contributes to the generation of this osmotic pressure (Koob et al., Bull. MDIBL 33, 5-8, 1994). These properties are similar to the physicochemical properties of the nucleus pulposus of the mammalian intervertebral disc, which derives in part from the embryonic notochord. The present report describes experiments on the isolated hagfish notochord examining the structural basis for this osmotic behavior and its contribution to specific biomechanical properties

Whole notochords were excised from propylene phenoxetol anesthetized hagfish (Myxine glutinosa). To measure osmotic properties, the core was removed by cutting open one end of the notochord and squeezing out the contents. Core tissue was then extracted with 6M guanidinium-HCl, 50 mM Na acetate, pH 6.5 for 24 hr at 4°C. The extracts were centrifuged at 27,000 x g for 30 min. The supernate was collected and dialyzed against 0.5 M NaCl, 1 mM NaH_2PO_4 for 24 hr at 4°C. One ml aliquots of the dialyzed extracts were placed in dialysis membranes. The membranes were blotted dry, weighed, and the contents were then dialyzed against 1 mM NaH_2PO_4, pH 7.0 containing NaCl concentrations varying between 0.15 M and 1.75 M for 24 hr, after which the dialysis bags were blotted and weighed again. Changes in wet weight of the dialyzed extracts greater than that of dialyzed control salt solutions are expressed as percentage change in wet weight compared to the weight of the initial 1 ml of extract.

Mechanical tests were performed on isolated notochord segments. Excised notochords were ligated with 00-silk suture at 4 cm intervals and transsected to produce uniform cylinders for tensile tests. Specimens were incubated in the neutral phosphate buffered salt solutions containing 0.15, 0.5 or 1.75 M NaCl for 24 hr at 4°C. Tensile tests were performed by attaching one end of the specimen with the suture to a fixed clamp. The other end was attached through the suture to a movable clamp mounted on nylon monofilament. The monofilament passed over a pulley and ended with a plastic beaker into which calibrated weights were added. Weights were added in 4 g increments, the tissue was allowed to elongate for 10 min at which time preliminary tests established it had reached equilibrium. Displacement was measured on graph paper mounted directly behind the beaker. Following the initial test, the suture at one end was carefully removed, the core tissue was squeezed out and replaced with an equivalent volume of the neutral salt buffer in which the original test was performed. The specimen was then tested as above.

When equilibrated in solutions of varying ionic strength, dialyzed notochord core guanidine extracts changed volume in inverse proportion to the external salt concentration (Fig. 1). Core extracts increased volume in NaCl concentrations below 0.5 M and decreased volume in NaCl above 0.5 M. These results are nearly identical to previous experiments on bulk swelling of whole notochord specimens (see Koob et al., op cit.), indicating that the swelling properties of the notochord are determined to a large extent by the physicochemical properties of the core.

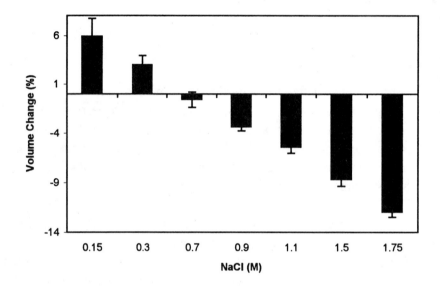

Figure 1. Volume change of notochord core extracts in dialysis tubing incubated for 24hr at 4°C in 1 mM NaH_2PO_4, pH 7.0, containing NaCl at the indicated concentrations. Volume change represents the relative change of weight over the starting weight. Values shown are means $\pm$ S.D. (n = 3/group).

Tensile tests revealed clear differences in the response of intact notochord segments to uniaxial load after equilibration in differing NaCl concentrations (Fig. 2A). The notochord was stiffest (i.e., least deformation per g load) when tested in 0.15 M NaCl. Specimens tested in 0.5 M NaCl, were less stiff than those tested in 0.15 M NaCl. Specimens tested in 1.5 M NaCl were significantly more extensible under load than those in the lower salt concentrations. When the core tissue was replaced with NaCl solutions equivalent to the original test solutions, the deformation/load curves of specimens tested in the three salt solutions were essentially identical (Fig. 2B).

These observations indicate that the physicochemical properties of the core tissue contribute significantly to the mechanical properties of the hagfish notochord. They suggest that, at the physiological osmolarity which is present in situ, the stiffness of the notochord and its ability to provide axial support are largely determined by the properties of the core macromolecules.

106

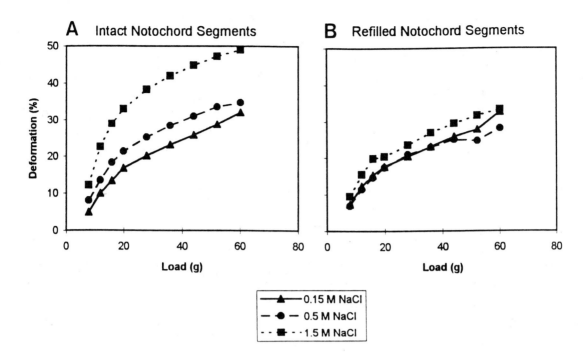

Figure 2. Tensile tests on isolated notochord segments with (A) and without (B) core tissue. Uniaxial tensile tests were performed in the indicated NaCl solutions containing 1 mM NaH$_2$PO$_4$, pH 7. Load was applied in 4 g increments, the specimen was allowed 10 min to reach equilibrium length after each addition of load. Deformation is expressed relative to the initial specimen length and values presented are means of three specimens.

Biochemical analyses of core extracts failed to detect proteoglycans (Koob et al., op. cit.) and we have been unable to find hyaluronic acid in the core, indicating that its composition is distinctly different from that in the mammalian nucleus pulposus. The principal protein found in extracts of the core is a molecule of apparent molecular weight of 45 kDa. It appears to be highly disulfide bonded since extraction done in the absence of reducing agents fails to solubilize it. Moreover, in order for the protein to enter a 4-20% linear SDS/PAGE gel, it must be reduced, suggesting that it exists in situ in large aggregates. We are currently characterizing this protein to determine the properties which contribute to the physicochemical properties of the notochord.

Funded by grants from the Shriners of North America to T.J.K. and Deutsche Forschungsgemeinschaft STO 71/9/1 to H.S.

ON THE HYDRODYNAMIC SHAPE OF LITTLE SKATE (RAJA ERINACEA) EGG CAPSULES

Thomas J. Koob[1] and Adam Summers[2]
[1]Skeletal Biology Section, Shriners Hospital, Tampa Unit, Tampa, FL 33612
[2]Organismic and Evolutionary Biology, University of Massachusetts, Amherst, MA 01003

While the unusual shape of skate egg capsules has evoked considerable speculation with respect to possible function, there has been no theoretical or empirical investigation of their shape-related properties. Little skate (Raja erinacea) egg capsules appear rectangular when viewed from the dorsal or ventral aspect; they are flattened in the dorsal-ventral plane with convex dorsal and ventral body walls (see Fig. 1). Four horns emanate from the corners and are oriented parallel with the long axis of the capsule. The anterior horns, curved ventrally, are shorter than the posterior horns, which bend towards one another. Slits on the outer seam of each horn, located about two thirds the distance to the end, are initially plugged with albumen. These slits open to admit sea water after approximately one third of the one to two year-long embryonic development. The present report describes flow tank experiments examining the hydrodynamic properties of these egg capsules, especially with regard to the influence of currents on flow of water through the capsule by way of the slits.

Little skate capsules at two stages of incubation were used for these experiments. Freshly laid capsules, collected within a week of oviposition, were used for measurement of drag. Freshly hatched capsules, laid in September 1994 and incubated in the year-round sea water system at the Laboratory, were used for experiments examining the effect of current on flow through the capsule, as well as for visualizing streamlines around the capsule. All experiments were performed in a laminar flow tank with a 5 x 1 x 1 ft working section. Current speed was measured with a video camera and video cassette recorder by introducing a bolus of dye and timing its transit through 20 cm. Five such measurements were made for each of five motor speeds at three separate locations in the working section of the tank. Speeds varied between 6 and 20 cm/sec.

Relative drag was measured by suspending capsules from a cantilever beam mounted on an axle and measuring the flow generated force exerted on the capsule through the beam to a calibrated Grass strain transducer. The output of a bridge amp connected to the strain gauge was amplified with a Tektronix DC amplifier and measured with an oscilloscope. Capsules were mounted in four orientations relative to the flow: the long axis parallel to flow, anterior horns forward; long axis parallel to flow, posterior horns forward; long axis perpendicular to flow, lateral seam facing forward; dorsal aspect facing flow. At each orientation, five current speeds between 6 and 20 cm/sec were tested in sequence from lowest to highest speed. Relative drag measurements were obtained for five randomly selected capsules.

To assess the effects of flow on water movement through the capsule, naturally hatched capsules were filled with methylthymol blue dissolved in sea water, the hatching seams were sealed with high pressure vacuum grease, and the capsules were placed on the floor of the tank and subjected to defined flow speeds for one hour. The contents of the capsules were then collected and their optical absorbance measured at 600 nm (λ_{max} for methylthymol blue at pH of MDIBL sea water). Methylthymol blue was chosen because preliminary experiments established that the body wall is impermeable to this dye, thus, loss of dye in these experiments could only occur through the slits. Capsules were subjected to five flow speeds ranging from 6 to 20 cm/sec. Loss of dye due to manipulation of the capsules as well as simple diffusion was measured by exposing dye-filled capsules to zero flow in the flow tank for one hour. Capsules were tested at three orientations relative to flow at 11 cm/sec for one hour: long axis parallel to flow, anterior horns forward; long axis parallel to flow, posterior horns forward; long axis perpendicular to flow. Six capsules were used for these experiments. The above experiments indicated that dye was leaving the capsules under flow. To determine through which horns water exited the capsule when exposed to flow at the various orientations, capsules were filled with milk and the flow of milk from the slits was recorded with a video camera.

Figure 1 illustrates the shape of little skate egg capsules. The midline sagittal section shows the streamline shape of the capsule. The fineness ratio (length/thickness) of 3.9 ± 0.4 (n = 21) and the position of the thickest dimension, one third of the distance behind the leading edge, are characteristic of streamlined bodies. The cross-section taken through the thickest portion of the capsule also shows considerable streamlining.

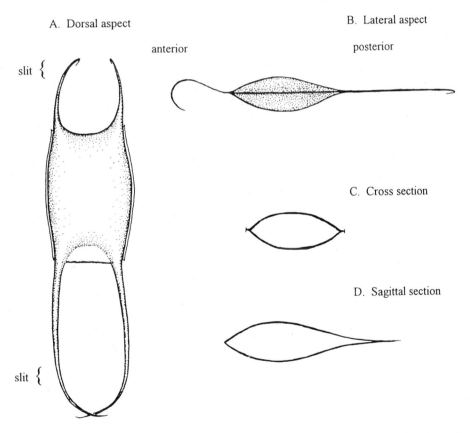

Figure 1. Illustration of the shape of little skate egg capsules. a) dorsal aspect; b) lateral aspect; c) cross section through thickest region.; d) midline sagittal section.

Relative drag measurements showed that the least drag occurred with the long axis of the capsule parallel to flow, regardless of which horns faced the current (Fig. 2). The capsule experienced somewhat higher drag when oriented perpendicular to flow with the lateral seam facing into the flow. Greatest drag occurred when the dorsal wall faced into the flow.

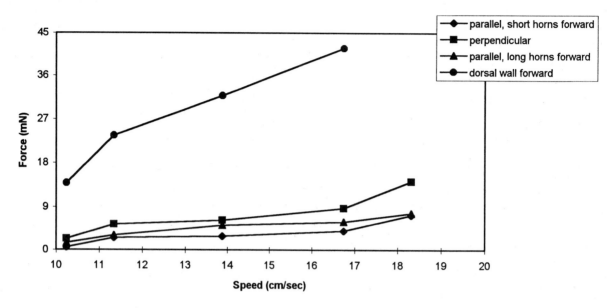

Figure 2. Relative drag measurements on capsules oriented as indicated at various flow speeds. Values presented are means for five capsules. Standard deviations are not shown because the values of the three curves at the bottom are not statistically different.

The effect of flow on the loss of dye from the capsules when oriented parallel to flow with the anterior horns facing forward is shown in Fig. 3. The amount of dye remaining in the capsule after one hour exposure to flow was directly related to flow speed above 6 cm/sec. A linear regression analysis using all values plus the starting optical absorbance of the dye gave a correlation coefficient R^2 of 0.78.

When tested at the three low drag orientations shown above, there was little difference in the amount of dye lost at 11 cm/sec flow. However, fluid exited the capsule through a particular set of horns depending on orientation. When placed with the long axis parallel to flow, regardless of which horns faced into the flow, milk exited the capsule only from the two downstream horns. When placed perpendicular to the flow, milk left through one downstream horn and the upstream horn diagonal to it. Further analyses will be necessary in order to determine the exact relationship between capsule orientation and flow induced water movement through particular sets of slits under various flow regimes. Coupled with pressure measurements, these experiments will provide an understanding of the precise factors that influence these hydrodynamic properties.

Little skate egg capsules exhibited specific hydrodynamic properties under the relatively ideal flow conditions examined here, including streamlining, low drag, and current induced flow

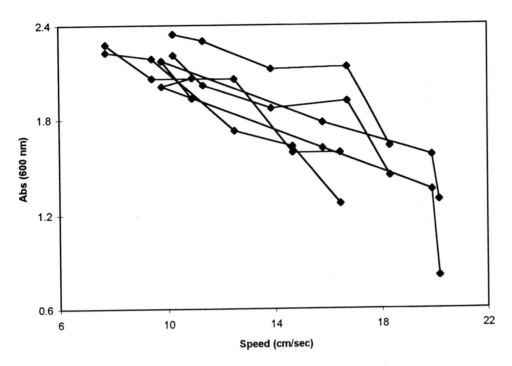

Figure 3. Scatter diagram of the optical absorbance of the methylthymol blue solutions remaining in six capsules after one hour at the indicated speeds. The lines connect the five values for individual capsules.

through the capsule via the slits. These hydrodynamic properties may enhance survivorship to hatching by providing a continual supply of fresh oxygen to satisfy the increasing respiratory demands as development progresses. However, while these capsules are clearly hydrodynamic, we can only speculate about the importance of their shape since natural spawning grounds have never been located. Therefore, the conditions, including flow patterns, under which these eggs successfully hatch are not known. As efforts to locate naturally deposited skate eggs continue, long-term experiments are underway to determine whether currents and current generated flow through the capsule are necessary for successful development to hatching.

The authors thank Peg Summers for assistance with the flow tank experiments and Wayne Mitchell for expert help in constructing the flow tank. Funded in part by the Shriners of North America #15961 to TJK, NSF IBN 9419892 to E.L. Brainerd, and the Investigators.

113

Class of 1999

Edward H, Benz, Jr., M.D.
William Osler Professor and Chair
Department of Medicine
Professor of Molecular Biology & Genetics
Johns Hopkins University School of Medicine

Thomas H. Maren, M.D.
Professor, Dept. of Pharmacology
University of Florida
College of Medicine

David C. Dawson, Ph.D.
Professor, Dept. of Physiology
University of Michigan Medical School

Mrs. Edith Milbury
Bar Harbor, Maine

Barbara Kent, Ph.D.
Administrative Director
Mount Desert Island Biological Lab

Raymond Rappaport, Ph.D.
Senior Research Scientist
Mount Desert Island Biological Lab

Mrs. Emily Leeser
Mount Desert, Maine
        and
New York, New York

Mrs. Edith Rudolf
Mount Desert, Maine
        and
New York, New York

SCIENTIFIC PERSONNEL 1995

Principal Investigator	Associates
Nazzareno Ballatori, Ph.D. Associate Professor Department of Environmental Medicine Environmental Health Science Center University of Rochester School of Medicine	A. Truong
Edward J. Benz, Jr., M.D. Professor and Chairman Department of Medicine Johns Hopkins University School of Medicine	J. Appel N. Yagoda
Nancy Berliner, M.D. Associate Professor of Medicine and Genetics Department of Internal Medicine/Hematology Section Yale University School of Medicine	M. Mulvey T. Wong
Marlies Betka, Ph.D. Research Associate Department of Biology Boston University	
James L. Boyer, M.D. Professor of Medicine Director, Liver Center Chief, Division of Digestive Diseases Yale University School of Medicine	A. Donald C. Fletcher D. Seward C. Soroka, Ph.D.
Gloria V. Callard, Ph.D. Professor, Department of Biology Boston University	L. McClusky
Ian P. Callard, Ph.D. Professor, Department of Biology Boston University	O. Putz
Horacio F. Cantiello, D.V.M., Ph.D. Asst. Professor of Medicine Harvard Medical School Assistant Biologist Massachusetts General Hospital	R. Jackson C. Jones A. Prat
James B. Claiborne, Ph.D. Professor, Department of Biology Georgia Southern University	J. Campbell L. Long
Lars Cleemann, Ph.D. Assistant Professor of Pharmacology Georgetown University	B. Arrue J. Fan

Principal Investigator	Associates
Gary Conrad, Ph.D. Professor, Division of Biology Kansas State University	A. Conrad, Ph.D. M. Janasek N. Martinez
Elizabeth L. Crockett, Ph.D. Assistant Professor Biological Sciences Ohio University	S. Vekasi E. Wilkes
David C. Dawson, Ph.D. Professor, Department of Physiology University of Michigan Medical School	B. Chen S. Cherukuri O. Omulepu J. Schafer
Franklin H. Epstein, M.D. William Applebaum Professor of Medicine, Harvard Medical School; Director, Nephrology Division, Beth Israel Hospital	K. Spokes
David H. Evans, Ph.D. Professor, Department of Zoology University of Florida	M. Gunderson
Bliss Forbush III, Ph.D. Professor of Cellular and Molecular Physiology Yale University School of Medicine	R. Behnke S. Houston A. Pernack
John N. Forrest, Jr., M.D. Professor, Department of Internal Medicine Yale University School of Medicine	S. Aller S. Bhanot G. Forrest A. George S. Kelmenson R. Lehrich C. Lubbe L. Matthews M. Ratner P. Schwartz
Raymond A. Frizzell, Ph.D. Prof. and Chairman, Dept. of Cell Biol/Phys. Univ. of Pittsburgh School of Medicine	A. Takahashi, Ph.D. E. Dawson
Leon Goldstein, Ph.D. Professor and Chairman Department of Physiology and Biophysics Brown University	A. Acosta E. Davis-Amaral M. Musch, Ph.D. E. Wilson

Principal Investigator	Associates
H. William Harris, M.D., Ph.D. Associate Professor, Dep. of Pediatrics Harvard Medical School; Director, Renal Research, Nephrology The Children's Hospital of Boston	
J.K. Haynes, Ph.D. David Packard Professor and Chairman Department of Biology Morehouse College	D. Sanderlin
John H. Henson, Ph.D. Assistant Professor Department of Biology Dickinson College	R. Mendola C. Roesner
Hartmut Hentschel, Ph.D. Senior Scientist Max-Planck Institut fuer Molekulare Physiologie Dortmund, Germany	M. Elger, Ph.D.
George W. Kidder III, Ph.D. Professor Department of Biological Sciences Illinois State University	P. Foster
Rolf Kinne, M.D., Ph.D. Director, Max-Planck Institut fuer Molekulare Physiologie Dortmund, Germany	C. Jette A. Werner
Evamaria Kinne-Saffran, M.D., Ph.D. Senior Investigator, Max-Planck Institut fuer Molekulare Physiologie Dortmund, Germany	A. Amstutz
Arnost Kleinzeller, M.D., Ph.D., D.Sc. Professor Emeritus Department of Physiology University of Pennsylvania	
Thomas J. Koob, Ph.D. Section Chief - Skeletal Biology Shriners Hospital for Crippled Children	S. Andrews M. Koob-Edmunds A. Summers J. Trotter, Ph.D.
Gregg A. Kormanik, Ph.D. Professor and Chairman Department of Biology University of North Carolina at Asheville	C. Harris

Principal Investigator	Associates
Thomas H. Maren, M.D. Graduate Research Professor Department of Pharmacology University of Florida College of Medicine	B. Taschner
James Maylie, Ph.D. Associate Professor Department of OB/GYN Oregon Health Sciences University	
Thomas J. McManus, M.D. Professor, Dept. of Cell Biology Duke University Medical Center	N. Vogenthaler L. Wood S. Kim
David S. Miller, Ph.D. Research Physiologist Laboratory of Cellular & Molecular Pharmacology NIH/NIEHS	L. Atherly
Martin Morad, Ph.D. Professor, Department of Pharmacology School of Medicine Georgetown University Medical Center	D. Harvey B. Hughes J. Monterrubio Y. Suzuki, Ph.D.
Alison Morrison-Shetlar, Ph.D. Professor, Biology Department Georgia Southern University	D. Soto B. Wolpin
Michael H. Nathanson, M.D., Ph.D. Associate Professor of Medicine and Cell Biology Liver Study Unit Yale University School of Medicine	K. Mariwalla
Carolyn R. Newton, Ph.D. Associate Professor Department of Biology Kalamazoo College	
Robert L. Preston, Ph.D. Professor of Physiology Department of Biological Sciences Illinois State University	C. Collins T. Gott J. Sharp
Raymond Rappaport, Ph.D. Senior Research Scientist Mount Desert Island Biological Laboratory	
J. Larry Renfro, Ph.D. Professor, Dept. of Physiology and Neurobiology University of Connecticut	

Principal Investigator	Associates
Carol E. Semrad, Ph.D. Asst. Professor of Clinical Medicine College of Physicians and Surgeons; Columbia University	
Charles Sidman, Ph.D. Professor Dept. of Molecular Genetics, Biochem, and Microbiology Univ. of Cincinatti College of Medicine	J. Anderson J. Rock S. Faull A. Wilson
Patricio Silva, M.D. Associate Professor, Dept. of Medicine, Harvard Medical School; Chief, Division of Nephrology New England Deaconess Hosp., Joslin Diabetes Center	R. Fraley D. Gumbs K. Mooney
Richard J. Solomon, M.D. Associate Professor of Medicine Harvard Medical School; Clinical Director, Division of Nephrology New England Deaconess Hospital	S. Solomon
Erik R. Swenson, M.D. Associate Professor of Medicine Pulmonary/Critical Care Division University of Washington	S. Vanderwerf
David W. Towle, Ph.D. Foster G. McGaw Professor Department of Biology Lake Forest College	M. Rushton J. Tilghman
Frances M. Van Dolah, Ph.D. Research Biochemist and Asst. Professor Charleston Laboratory; U.S. National Marine Fisheries Service; Marine Biomedical and Environmental Sciences; Medical University of South Carolina	T. Leighfield L. Sugg
Rui Wang, M.D., Ph.D. Assistant Professor Department of Physiology University of Montreal, Canada	M. Branch A. Doyle
Jose A. Zadunaisky, M.D., Ph.D. Professor of Physiology and Biophysics Professor of Experimental Ophthalmology New York University Medical Center	M. Balla D. Croft

1 9 9 5   S E M I N A R S

Morning Transport

July 10   "CFTR: Searching for the Pore" David C. Dawson, Ph.D., University of Michigan Medical School; Director, MDIBL

July 17   "An ATP-sensitive volume regulatory osmolyte channel" Ned Ballatori, Ph.D., University of Rochester School of Medicine

July 24   "Stoichiometry of voltage-dependent K^+-channel, min K" James Maylie, Ph.D., Oregon Health Sciences University

July 31   "Heat-shock proteins" J. Larry Renfro, Ph.D., The University of Connecticut

August 7   "Ca^{++} transport systems as evolutionary markers: In the beginning there was the Na^+/Ca^{++} exchanger" Martin Morad, Ph.D., Georgetown University Medical Center

August 14 "Cloning of the C-type natriuretic hormone receptor from the shark rectal gland" John N. Forrest, Jr., M.D., Yale University School of Medicine

August 21 "Role of cytoskeleton in CFTR regulation" Horacio F. Cantiello, D.V.M., Ph.D., Harvard Medical School; Massachusetts General Hospital

Noon

June 30 and July 7 - MDIBL Principal Investigators presented a series of 5 minute talks to outline summer research plans

July 21   "Mechanism of cholera toxin stimulation of chloride secretion" David Burleigh, Ph.D., Lecturer, University of London

July 28   "The role of cholesterol in the thermal adaptation of plasma membranes from poikilotherms" Elizabeth Crockett, Ph.D., Ohio University

"Pathways stimulating rectal gland chloride secretion" Kathrina Mooney, Oglethorpe University

August 4   "Role of Ca channels in chromaffin cell secretion" Lars Cleemann, Ph.D., Georgetown University Medical Center

"ATP receptors in skate hepatocytes" Michael H. Nathanson, M.D., Ph.D., Yale University School of Medicine

Evening

July 5   "Endocrine disrupting environmental contaminants and embryos: Lessons from wildlife" Louis J. Guillette, Jr., Ph.D., University of Florida

July 12   "Toxin production and growth regulation in marine 'Red Tide' dino-
           flagellates" Frances M. Van Dolah, Ph.D., Marine Biomedical and
           Environmental Sciences, Medical University of South Carolina

July 19   THE FIFTH THOMAS H. MAREN LECTURE - "K channels and K secretion in
           the collecting duct" Lawrence G. Palmer, Ph.D., Cornell University
           Medical College, New York

July 26   "Multiple site optical recording of electrical activity: Designer-
           built 2-dimensional hearts and a mammalian simple nervous system"
           Brian M. Salzberg, Ph.D., University of Pennsylvania School of
           Medicine

August 2   THE FOURTH JOHN W. BOYLAN MEMORIAL LECTURE - "North American estu-
           aries and coastal waters: Passageways to the future" Robert H.
           Boyle, M.A. (Yale University). Reporter, Staff Writer, Staff Corre-
           spondent, Sports Illustrated and Time, 1954-1960; Associate Edi-
           tor, Senior Editor and Senior Writer, Sports Illustrated, 1960-
           1986; Special Contributor, SI, 1986-present.

August 9   THE FIRST HELEN F. CSERR MEMORIAL LECTURE - "Downstream Interactors
           of MIS, Activin, and TGF-β Type I Receptor" Patricia K. Donahoe,
           M.D., Chief, Pediatric Surgery, Massachusetts General Hospital

August 23 "Coordinate regulation of neutrophil secondary granule protein gene
           expression" Nancy Berliner, M.D., Yale University School of Medi-
           cine

Special Seminars

June 22   "Living Fossils Around the World" Jennifer Rock, B.A., College of
           the Atlantic

July 6   "Hemoglobin and hemocyanin expression: The effects of time and
           tides" Nora Terwilliger, Ph.D., Oregon Institute of Marine Biology

August 3   All campus MDIBL ethics seminar - "Approaches to environmental
           ethics" Roger King, Ph.D., Department of Philosophy, University of
           Maine. Compulsory for all fellowship recipients.

August 3   "Future Science: Scientists of Tomorrow at Work Today" Public
           seminar featuring presentations by local and national high school
           and college students participating in exciting hands-on research
           training programs at MDIBL and the Jackson Laboratory

                   STUDENTS' "BASIC" LECTURE SERIES
                 (Participation required by students who
                 are recipients of fellowships/scholarships)

June 27   "Actin reorganization in response to cell volume changes" Colleen
           Roesener, Dickinson College

July 11   1) "MDIBL: the last 15 years"; 2) "Two unlikely endocrine tis-
           sues: The heart and the endotheliate lining of blood vessels"
           David H. Evans, Ph.D., University of Florida and Mark Gunderson,
           St. Olaf College

July 18   "Transport in cell membranes of the eye and gill" Jose Zadunaisky, M.D., Ph.D., and David Croft, New York University Medical School

July 25   "Evolution of life histories - what does physiology tell us?" Oliver Putz, Boston University

August 1   "Renal excretion of toxic chemicals - How your kidney tells drugs & pollutants to piss-off!" David S. Miller, Ph.D., NIEHS

August 8   "Ca^{++} signaling in the skate liver: Experimental methods for the TV generation" Kavita Mariwalla, Yale University

            "Technician information session" Dave Opdyke, Ursinus College, and Justin Crouse, Wesleyan University

August 14   "Cystic fibrosis: Mutations, mechanisms and models" David C. Dawson, Ph.D., University of Michigan Medical School; Director, MDIBL

Symposia

August 10   The Second Annual Maine Symposium on Molecular Toxicology sponsored by the Maine Toxicology Institute/Eastern Maine Medical Center; Co-host, MDIBL Center for Membrane Toxicity Studies - "Cellular Defense Against Toxic Chemicals: Extrusion Via Specific Pumps and Transporters", Moderator, James L. Boyer, Director, Center for Membrane Toxicity Studies

            "Overview: Glutathione S-Conjugate Transporters and Pumps" Ned Ballatori, Ph.D., University of Rochester School of Medicine

            "Metalloregulation of an Arsenite (Antimonite)-Translocating AT-Pase" Barry P. Rosen, Ph.D., Wayne State University School of Medicine

            "Structure/Function Studies on the Wilson and Menkes Disease P-Type ATPases" Jonathan D. Gitlin, M.D., Washington University School of Medicine

            "MRP, An Alternative Cause of Multidrug Resistance" Roger G. Deeley, Ph.D., Cancer Research Laboratories, Queen's University

            "Electrophysiology of P-Glycoprotein: The Missing Link?" Horacio Cantiello, D.V.M., Ph.D., Harvard University School of Medicine

            "P-Glycoprotein Expression in Yeast" Philippe Gros, Ph.D., McGill University

            THE FOURTEENTH WILLIAM B. KINTER MEMORIAL LECTURE - "Mechanism and Manipulation of the Multidrug Transporter" Michael Gottesman, M.D., Laboratory of Cell Biology, National Cancer Institute, NIH

August 15-19, 1995 - AN INTERNATIONAL SYMPOSIUM HONORING HOMER W. SMITH ON THE 100TH ANNIVERSARY OF HIS BIRTH - "The Kidney: Structure and Function in Health and Disease"

1 9 9 5   P U B L I C A T I O N S

Ballatori, N., A.T. Truong, P.S. Jackson, K. Strange and J.L. Boyer. ATP
depletion and inactivation of an ATP-sensitive taurine channel by
classic ion channel blockers. Mol. Pharmacol. 48:472-476, 1995.

Burns, A., C. Roesener, J. Zmuda, R. Mendola, and J. Henson. Actin-based
cortical flow: regulation by intracellular ionic conditions and involvement in
cellular wound resealing. Molecular Biology of the Cell 6:20a, 1995.

Evans, D.H. The roles of natriuretic peptide hormones (NPs) in fish osmoregulation. In: Advances in Environmental and Comparative Physiology--Mechanisms
of Systemic Regulation in Lower Vertebrates II: Acid-base Regulation, Ion
Transfer and Metabolism. N. Heisler, ed. Springer-Verlag, Heidelberg, pp. 119-
152, 1995.

Evans, D.H., M. Gunderson, and C. Cegelis. ETβ-type receptors mediate en-
dothelin-stimulated contraction in the aortic vascular smooth muscle of the
spiny dogfish shark, Squalus acanthias. J. Comp. Physiol. (in press, 1996).

Henson, J.H., S. Capuano, D. Nesbitt, D.N. Hager, S. Nundy, D.S. Miller, N.
Ballatori and J.L. Boyer. Cytoskeletal organization in clusters of isolated
polarized skate hepatocytes: Structural and functional evidence for microtubule-dependent transcytosis. J. Exp. Zool. 271:273-284, 1995.

Henson, J.H., D.G. Cole, M. Terasaki, D. Rashid, and J.M. Scholey. Immunola-
calization of the heterotrimeric kinesin related protein KRP (85/95) to the
mitotic apparatus of early sea urchin embryos. Developmental Biology 171:182-
194, 1995.

Kidder, G.W. III, and A.A. McCoy. The electromyogram as a measure of heavy
metal toxicity in fresh water and salt water mussels. (Bull. Environ. Contam.
Toxicol., in press.

Kormanik, G.A. Maternal-fetal transfer of nitrogen in chondrichthians. In:
Nitrogen Metabolism and Excretion. P. Walsh and P. Wright, eds. CRC Press,
Boca Raton, FL, 1995.

Maylie, J. and M. Morad. Evaluation of T- and L-type Ca^{2+} currents in shark
ventricular myocytes. Am. J. Physiol, H1695-H1704, 1995.

Miller, D.S., G. Fricker, J.H. Henson, D.N. Hager, S. Nundy, N. Ballatori, and
J.L. Boyer. Active, microtubule-dependent secretion of a fluorescent bile
salt derivative in skate hepatocyte clusters. American Journal of Physiology,
in press.

Nathanson, M.H., K. Mariwalla, N. Ballatori, and J.L. Boyer. Effects of Hg^{2+}
on cytosolic Ca^{2+} in isolated skate hepatocytes. Cell Calcium 18:429-439,
1995.

Nathanson, M.H. and K. Mariwalla. Characterization and function of ATP receptors on hepatocytes from the little skate, Raja erinacea. American Journal of
Physiology: Regulatory, Integrative and Comparative Physiology, (in press).

Schramm, U., G. Fricker, R. Wenger, and D.S. Miller. $_p$-Glycoprotein-mediated secretion of a fluorescent cyclosporin analogue by teleost renal proximal tubules. Am. J. Physiol., 268:F46-F52, 1995.

Toop, T., J.A. Donald, and D.H. Evans. Localisation and characteristics of natriuretic peptide receptors in the gills of the Atlantic hagfish, Myxine glutinosa (Agnatha). J. Exp. Biol. 198:117-126, 1995.

Toop, T., J.A. Donald, and D.H. Evans. Natriuretic peptide receptors in the kidney and the ventral and dorsal aortae of the Atlantic hagfish, Myxine glutinosa (Agnatha). J. Exp. Biol. 198:1875-1882, 1995.

AUTHOR INDEX

S P E C I E S   I N D E X